www.EffortlessMath.com

... So Much More Online!

✓ FREE Math lessons

✓ More Math learning books!

✓ Mathematics Worksheets

✓ Online Math Tutors

Need a PDF version of this book?

Please visit www.EffortlessMath.com

Comprehensive SIFT Math Practice Book 2020 - 2021

Complete Coverage of all SIFT Math Concepts + 2 Full-Length SIFT Math Tests

By

Reza Nazari & Ava Ross

Copyright © 2020

Reza Nazari & Ava Ross

All rights reserved. No part of this publication may be reproduced, stored in a retrieval system, or transmitted in any form or by any means, electronic, mechanical, photocopying, recording, scanning, or otherwise, except as permitted under Section 107 or 108 of the 1976 United States Copyright Ac, without permission of the author.

All inquiries should be addressed to:

info@effortlessMath.com

www.EffortlessMath.com

ISBN: 978-1-64612-474-9

Published by: Effortless Math Education

www.EffortlessMath.com

Visit www.EffortlessMath.com
for Online Math Practice

Description

Comprehensive SIFT Math Practice Book 2020 - 2021, which reflects the 2020 - 2021 test guidelines, is a precious learning resource for SIFT test-takers who need extra practice in math to raise their SIFT Math scores. Upon completion of this exercise book, you will have a solid foundation and sufficient practice to ace the SIFT Math test. **This comprehensive practice book is your ticket to scoring higher on SIFT Math.**

The updated version of this unique practice workbook represents extensive exercises, math problems, sample SIFT questions, and quizzes with answers and detailed solutions to help you hone your math skills, overcome your exam anxiety, boost your confidence—and do your best to defeat the SIFT exam on test day.

Comprehensive SIFT Math Practice Book 2020 – 2021 includes many exciting and unique features to help you improve your test scores, including:

- ✓ Content 100% aligned with the 2020 SIFT Math test
- ✓ Complete coverage of all SIFT Math concepts and topics which you will be tested
- ✓ Over 2,500 additional SIFT math practice questions in both multiple-choice and grid-in formats with answers grouped by topic, so you can focus on your weak areas
- ✓ Abundant Math skill-building exercises to help test-takers approach different question types that might be unfamiliar to them
- ✓ 2 full-length practice tests (featuring new question types) with detailed answers

This SIFT Math practice book and other Effortless Math Education books are used by thousands of students each year to help them review core content areas, brush-up in math, discover their strengths and weaknesses, and achieve their best scores on the SIFT test.

Contents

Description 4

Chapter 1: Fractions and Mixed Numbers 9
Simplifying Fractions 10
Adding and Subtracting Fractions 11
Multiplying and Dividing Fractions 12
Adding Mixed Numbers 13
Subtracting Mixed Numbers 14
Multiplying Mixed Numbers 15
Dividing Mixed Numbers 16
Answers – Chapter 1 17

Chapter 2: Decimals 20
Comparing Decimals 21
Rounding Decimals 22
Adding and Subtracting Decimals 23
Multiplying and Dividing Decimals 24
Answers – Chapter 2 25

Chapter 3: Integers and Order of Operations 27
Adding and Subtracting Integers 28
Multiplying and Dividing Integers 29
Order of Operation 30
Integers and Absolute Value 31
Answers – Chapter 3 32

Chapter 4: Ratios and Proportions 34
Simplifying Ratios 35
Proportional Ratios 36
Create Proportion 37
Similarity and Ratios 38
Simple Interest 39
Answers – Chapter 4 40

Chapter 5: Percentage 42

Percent Problems 43

Percent of Increase and Decrease 44

Discount, Tax and Tip 45

Answers – Chapter 5 46

Chapter 6: Expressions and Variables 47

Simplifying Variable Expressions 48

Simplifying Polynomial Expressions 49

Evaluating One Variable 50

Evaluating Two Variables 51

The Distributive Property 52

Answers – Chapter 6 53

Chapter 7: Equations and Inequalities 55

One–Step Equations 56

Multi –Step Equations 57

System of Equations 58

Graphing Single–Variable Inequalities 59

One–Step Inequalities 60

Multi –Step Inequalities 61

Answers – Chapter 7 62

Chapter 8: Lines and Slope 65

Finding Slope 66

Graphing Lines Using Slope–Intercept Form 67

Writing Linear Equations 68

Finding Midpoint 69

Finding Distance of Two Points 70

Answers – Chapter 8 71

Chapter 9: Exponents and Variables 73

Multiplication Property of Exponents 74

Division Property of Exponents 75

Powers of Products and Quotients 76

Zero and Negative Exponents 77

Negative Exponents and Negative Bases 78

Scientific Notation 79

Radicals 80

Answers – Chapter 9 81

Chapter 10: Polynomials 84

Simplifying Polynomials 85

Adding and Subtracting Polynomials 86

Multiplying Monomials 87

Multiplying and Dividing Monomials 88

Multiplying a Polynomial and a Monomial 89

Multiplying Binomials 90

Factoring Trinomials 91

Answers – Chapter 10 92

Chapter 11: Geometry and Solid Figures 94

The Pythagorean Theorem 95

Triangles 96

Polygons 97

Circles 98

Cubes 99

Trapezoids 100

Rectangular Prisms 101

Cylinder 102

Answers – Chapter 11 103

Chapter 12: Statistics 105

Mean, Median, Mode, and Range of the Given Data 106

Pie Graph 107

Probability Problems 108

Permutations and Combinations 109

Answers – Chapter 12 110

Chapter 13: Functions Operations 111

Function Notation and Evaluation 112

Adding and Subtracting Functions 113

Multiplying and Dividing Functions 114

Composition of Functions 115

Answers – Chapter 13 116

www.EffortlessMath.com

Chapter 14: Quadratic..117
    Solving a Quadratic Equation...118
    Graphing Quadratic Functions 119
    Solving Quadratic Inequalities 120
    Graphing Quadratic Inequalities 121
    Answers – Chapter 14 122
SIFT Test Review..124
SIFT Math Practice Test 1..126
SIFT Math Practice Test 2 137
SIFT Math Practice Tests Answer Keys 148
SIFT Math Practice Tests Answers and Explanations.................................149

Chapter 1:
Fractions and Mixed Numbers

Math Topics that you'll learn in this Chapter:

- ✓ Simplifying Fractions
- ✓ Adding and Subtracting Fractions
- ✓ Multiplying and Dividing Fractions
- ✓ Adding Mixed Numbers
- ✓ Subtracting Mixed Numbers
- ✓ Multiplying Mixed Numbers
- ✓ Dividing Mixed Numbers

Simplifying Fractions

✎ *Simplify each fraction.*

1) $\frac{10}{15} =$

2) $\frac{8}{20} =$

3) $\frac{12}{42} =$

4) $\frac{5}{20} =$

5) $\frac{6}{18} =$

6) $\frac{18}{27} =$

7) $\frac{15}{55} =$

8) $\frac{24}{54} =$

9) $\frac{63}{72} =$

10) $\frac{40}{64} =$

11) $\frac{23}{46} =$

12) $\frac{35}{63} =$

13) $\frac{32}{36} =$

14) $\frac{81}{99} =$

15) $\frac{16}{64} =$

16) $\frac{14}{35} =$

17) $\frac{19}{38} =$

18) $\frac{18}{54} =$

19) $\frac{56}{70} =$

20) $\frac{40}{45} =$

21) $\frac{9}{90} =$

22) $\frac{20}{25} =$

23) $\frac{32}{48} =$

24) $\frac{7}{49} =$

25) $\frac{18}{48} =$

26) $\frac{54}{108} =$

Adding and Subtracting Fractions

✎ *Calculate and write the answer in lowest term.*

1) $\dfrac{1}{5} + \dfrac{1}{7} =$

2) $\dfrac{3}{7} + \dfrac{4}{5} =$

3) $\dfrac{3}{8} - \dfrac{1}{9} =$

4) $\dfrac{4}{5} - \dfrac{5}{9} =$

5) $\dfrac{2}{9} + \dfrac{1}{3} =$

6) $\dfrac{3}{10} + \dfrac{2}{5} =$

7) $\dfrac{9}{10} - \dfrac{4}{5} =$

8) $\dfrac{7}{9} - \dfrac{3}{7} =$

9) $\dfrac{3}{4} + \dfrac{1}{3} =$

10) $\dfrac{3}{8} + \dfrac{2}{5} =$

11) $\dfrac{3}{4} - \dfrac{2}{5} =$

12) $\dfrac{7}{9} - \dfrac{2}{3} =$

13) $\dfrac{4}{9} + \dfrac{5}{6} =$

14) $\dfrac{2}{3} + \dfrac{1}{4} =$

15) $\dfrac{9}{10} - \dfrac{3}{5} =$

16) $\dfrac{7}{12} - \dfrac{1}{2} =$

17) $\dfrac{4}{5} + \dfrac{2}{3} =$

18) $\dfrac{5}{7} + \dfrac{1}{5} =$

19) $\dfrac{5}{9} - \dfrac{2}{5} =$

20) $\dfrac{3}{5} - \dfrac{2}{9} =$

21) $\dfrac{7}{9} + \dfrac{1}{7} =$

22) $\dfrac{5}{8} + \dfrac{2}{3} =$

23) $\dfrac{4}{7} + \dfrac{2}{3} =$

24) $\dfrac{6}{7} - \dfrac{4}{9} =$

25) $\dfrac{4}{5} - \dfrac{2}{15} =$

26) $\dfrac{2}{9} + \dfrac{4}{5} =$

Multiplying and Dividing Fractions

✎ *Solve and write the answer in lowest term.*

1) $\frac{1}{2} \times \frac{4}{5} =$

2) $\frac{1}{5} \times \frac{6}{7} =$

3) $\frac{1}{3} \div \frac{1}{7} =$

4) $\frac{1}{7} \div \frac{3}{8} =$

5) $\frac{2}{3} \times \frac{4}{7} =$

6) $\frac{5}{7} \times \frac{3}{4} =$

7) $\frac{2}{5} \div \frac{3}{7} =$

8) $\frac{3}{7} \div \frac{5}{8} =$

9) $\frac{3}{8} \times \frac{4}{7} =$

10) $\frac{2}{9} \times \frac{6}{11} =$

11) $\frac{1}{10} \div \frac{3}{8} =$

12) $\frac{3}{10} \div \frac{4}{5} =$

13) $\frac{6}{7} \times \frac{4}{9} =$

14) $\frac{3}{7} \times \frac{5}{6} =$

15) $\frac{7}{9} \div \frac{6}{11} =$

16) $\frac{1}{15} \div \frac{2}{3} =$

17) $\frac{1}{13} \times \frac{1}{2} =$

18) $\frac{1}{12} \times \frac{4}{7} =$

19) $\frac{1}{15} \div \frac{4}{9} =$

20) $\frac{1}{16} \div \frac{1}{2} =$

21) $\frac{4}{7} \times \frac{5}{8} =$

22) $\frac{1}{11} \times \frac{4}{5} =$

23) $\frac{1}{18} \div \frac{5}{6} =$

24) $\frac{1}{15} \div \frac{3}{8} =$

25) $\frac{1}{11} \times \frac{3}{4} =$

26) $\frac{1}{14} \times \frac{2}{3} =$

Adding Mixed Numbers

✎ *Solve and write the answer in lowest terms.*

1) $3\frac{1}{5} + 2\frac{2}{9} =$

2) $1\frac{1}{7} + 5\frac{2}{5} =$

3) $4\frac{4}{5} + 1\frac{2}{7} =$

4) $2\frac{4}{7} + 2\frac{3}{5} =$

5) $1\frac{5}{6} + 1\frac{2}{5} =$

6) $3\frac{5}{7} + 1\frac{2}{9} =$

7) $3\frac{5}{8} + 2\frac{1}{3} =$

8) $1\frac{6}{7} + 3\frac{2}{9} =$

9) $2\frac{5}{9} + 1\frac{1}{4} =$

10) $3\frac{7}{9} + 2\frac{5}{6} =$

11) $2\frac{1}{10} + 2\frac{2}{5} =$

12) $1\frac{3}{10} + 3\frac{4}{5} =$

13) $3\frac{1}{12} + 2\frac{1}{3} =$

14) $5\frac{1}{11} + 1\frac{1}{2} =$

15) $3\frac{1}{21} + 2\frac{2}{3} =$

16) $4\frac{1}{24} + 1\frac{5}{8} =$

17) $2\frac{1}{25} + 3\frac{3}{5} =$

18) $3\frac{1}{15} + 2\frac{2}{10} =$

19) $5\frac{6}{7} + 2\frac{1}{3} =$

20) $2\frac{1}{8} + 3\frac{3}{4} =$

21) $2\frac{5}{7} + 2\frac{2}{21} =$

22) $4\frac{1}{6} + 1\frac{4}{5} =$

23) $3\frac{5}{6} + 1\frac{2}{7} =$

24) $2\frac{7}{8} + 3\frac{1}{3} =$

25) $3\frac{1}{17} + 1\frac{1}{2} =$

26) $1\frac{1}{18} + 1\frac{4}{9} =$

www.EffortlessMath.com

Subtracting Mixed Numbers

✎ *Solve and write the answer in lowest terms.*

1) $3\frac{2}{5} - 1\frac{2}{9} =$

2) $5\frac{3}{5} - 1\frac{1}{7} =$

3) $4\frac{2}{5} - 2\frac{2}{7} =$

4) $8\frac{3}{4} - 2\frac{1}{8} =$

5) $9\frac{5}{7} - 7\frac{4}{21} =$

6) $11\frac{7}{12} - 9\frac{5}{6} =$

7) $9\frac{5}{9} - 8\frac{1}{8} =$

8) $13\frac{7}{9} - 11\frac{3}{7} =$

9) $8\frac{7}{12} - 7\frac{3}{8} =$

10) $11\frac{5}{9} - 9\frac{1}{4} =$

11) $6\frac{5}{6} - 2\frac{2}{9} =$

12) $5\frac{7}{8} - 4\frac{1}{3} =$

13) $9\frac{5}{8} - 8\frac{1}{2} =$

14) $4\frac{9}{16} - 2\frac{1}{4} =$

15) $3\frac{2}{3} - 1\frac{2}{15} =$

16) $5\frac{1}{2} - 4\frac{2}{17} =$

17) $5\frac{6}{7} - 2\frac{1}{3} =$

18) $3\frac{3}{7} - 2\frac{2}{21} =$

19) $7\frac{3}{10} - 5\frac{2}{15} =$

20) $4\frac{5}{6} - 2\frac{2}{9} =$

21) $6\frac{3}{7} - 2\frac{2}{9} =$

22) $7\frac{4}{5} - 6\frac{3}{7} =$

23) $10\frac{2}{3} - 9\frac{5}{8} =$

24) $9\frac{3}{4} - 7\frac{4}{9} =$

25) $15\frac{4}{5} - 13\frac{12}{25} =$

26) $13\frac{5}{12} - 7\frac{5}{24} =$

Multiplying Mixed Numbers

✎ *Solve and write the answer in lowest terms.*

1) $1\frac{1}{8} \times 1\frac{3}{4} =$

2) $3\frac{1}{5} \times 2\frac{2}{7} =$

3) $2\frac{1}{8} \times 1\frac{2}{9} =$

4) $2\frac{3}{8} \times 2\frac{2}{5} =$

5) $1\frac{1}{2} \times 5\frac{2}{3} =$

6) $3\frac{1}{2} \times 6\frac{2}{3} =$

7) $9\frac{1}{2} \times 2\frac{1}{6} =$

8) $2\frac{5}{8} \times 8\frac{3}{5} =$

9) $3\frac{4}{5} \times 4\frac{2}{3} =$

10) $5\frac{1}{3} \times 2\frac{2}{7} =$

11) $6\frac{1}{3} \times 3\frac{3}{4} =$

12) $7\frac{2}{3} \times 1\frac{8}{9} =$

13) $8\frac{1}{2} \times 2\frac{1}{6} =$

14) $4\frac{1}{5} \times 8\frac{2}{3} =$

15) $3\frac{1}{8} \times 5\frac{2}{3} =$

16) $2\frac{2}{7} \times 6\frac{2}{5} =$

17) $2\frac{3}{8} \times 7\frac{2}{3} =$

18) $1\frac{7}{8} \times 8\frac{2}{3} =$

19) $9\frac{1}{2} \times 3\frac{1}{5} =$

20) $2\frac{5}{8} \times 4\frac{1}{3} =$

21) $6\frac{1}{3} \times 3\frac{2}{5} =$

22) $5\frac{3}{4} \times 2\frac{2}{7} =$

23) $9\frac{1}{4} \times 2\frac{1}{3} =$

24) $3\frac{3}{7} \times 7\frac{2}{5} =$

25) $4\frac{1}{4} \times 3\frac{2}{5} =$

26) $7\frac{2}{3} \times 3\frac{2}{5} =$

Dividing Mixed Numbers

Solve and write the answer in lowest terms.

1) $9\frac{1}{2} \div 2\frac{3}{5} =$

2) $2\frac{3}{8} \div 1\frac{2}{5} =$

3) $5\frac{3}{4} \div 2\frac{2}{7} =$

4) $8\frac{1}{3} \div 4\frac{1}{4} =$

5) $7\frac{2}{5} \div 3\frac{3}{4} =$

6) $2\frac{4}{5} \div 3\frac{2}{3} =$

7) $8\frac{3}{5} \div 4\frac{3}{4} =$

8) $6\frac{3}{4} \div 2\frac{2}{9} =$

9) $5\frac{2}{7} \div 2\frac{2}{9} =$

10) $2\frac{2}{5} \div 3\frac{3}{5} =$

11) $4\frac{3}{7} \div 1\frac{7}{8} =$

12) $2\frac{5}{7} \div 2\frac{4}{5} =$

13) $8\frac{3}{5} \div 6\frac{1}{5} =$

14) $2\frac{5}{8} \div 1\frac{8}{9} =$

15) $5\frac{6}{7} \div 2\frac{3}{4} =$

16) $1\frac{3}{5} \div 2\frac{3}{8} =$

17) $5\frac{3}{4} \div 3\frac{2}{5} =$

18) $2\frac{3}{4} \div 3\frac{1}{5} =$

19) $3\frac{2}{3} \div 1\frac{2}{5} =$

20) $4\frac{1}{4} \div 2\frac{2}{3} =$

21) $3\frac{5}{6} \div 2\frac{4}{5} =$

22) $2\frac{1}{8} \div 1\frac{3}{4} =$

23) $5\frac{1}{2} \div 2\frac{2}{5} =$

24) $3\frac{4}{7} \div 2\frac{2}{3} =$

25) $2\frac{4}{5} \div 3\frac{5}{6} =$

26) $2\frac{3}{7} \div 3\frac{2}{3} =$

Answers – Chapter 1

Simplifying Fractions

1) $\frac{2}{3}$
2) $\frac{2}{5}$
3) $\frac{2}{7}$
4) $\frac{1}{4}$
5) $\frac{1}{3}$
6) $\frac{2}{3}$
7) $\frac{3}{11}$
8) $\frac{4}{9}$
9) $\frac{7}{8}$
10) $\frac{5}{8}$
11) $\frac{1}{2}$
12) $\frac{5}{9}$
13) $\frac{8}{9}$
14) $\frac{9}{11}$
15) $\frac{1}{4}$
16) $\frac{2}{5}$
17) $\frac{1}{2}$
18) $\frac{1}{3}$
19) $\frac{4}{5}$
20) $\frac{8}{9}$
21) $\frac{1}{10}$
22) $\frac{4}{5}$
23) $\frac{2}{3}$
24) $\frac{1}{7}$
25) $\frac{3}{8}$
26) $\frac{1}{2}$

Adding and Subtracting Fractions

1) $\frac{12}{35}$
2) $\frac{43}{35}$
3) $\frac{19}{72}$
4) $\frac{11}{45}$
5) $\frac{5}{9}$
6) $\frac{7}{10}$
7) $\frac{1}{10}$
8) $\frac{22}{63}$
9) $\frac{13}{12}$
10) $\frac{31}{40}$
11) $\frac{7}{20}$
12) $\frac{1}{9}$
13) $\frac{23}{18}$
14) $\frac{11}{12}$
15) $\frac{3}{10}$
16) $\frac{1}{12}$
17) $\frac{22}{15}$
18) $\frac{32}{35}$
19) $\frac{7}{45}$
20) $\frac{17}{45}$
21) $\frac{58}{63}$
22) $\frac{31}{24}$
23) $\frac{26}{21}$
24) $\frac{26}{63}$
25) $\frac{2}{3}$
26) $\frac{46}{45}$

Multiplying and Dividing Fractions

1) $\frac{2}{5}$
2) $\frac{6}{35}$
3) $\frac{7}{3}$
4) $\frac{8}{21}$
5) $\frac{8}{21}$
6) $\frac{15}{28}$
7) $\frac{14}{15}$
8) $\frac{24}{35}$
9) $\frac{3}{14}$
10) $\frac{4}{33}$
11) $\frac{4}{15}$
12) $\frac{3}{8}$
13) $\frac{8}{21}$
14) $\frac{5}{14}$
15) $\frac{77}{54}$
16) $\frac{1}{10}$
17) $\frac{1}{26}$
18) $\frac{1}{21}$
19) $\frac{3}{20}$
20) $\frac{1}{8}$

21) $\frac{5}{14}$ 23) $\frac{1}{15}$ 25) $\frac{3}{44}$

22) $\frac{4}{55}$ 24) $\frac{8}{45}$ 26) $\frac{1}{21}$

Adding Mixed Numbers

1) $5\frac{19}{45}$ 8) $5\frac{5}{63}$ 15) $5\frac{5}{7}$ 22) $5\frac{29}{30}$

2) $6\frac{19}{35}$ 9) $3\frac{29}{36}$ 16) $5\frac{2}{3}$ 23) $5\frac{5}{42}$

3) $6\frac{3}{35}$ 10) $6\frac{11}{18}$ 17) $5\frac{16}{25}$ 24) $6\frac{5}{24}$

4) $5\frac{6}{35}$ 11) $4\frac{1}{2}$ 18) $5\frac{4}{15}$ 25) $4\frac{19}{34}$

5) $3\frac{7}{30}$ 12) $5\frac{1}{10}$ 19) $8\frac{4}{21}$ 26) $2\frac{1}{2}$

6) $4\frac{59}{63}$ 13) $5\frac{5}{12}$ 20) $5\frac{7}{8}$

7) $5\frac{23}{24}$ 14) $6\frac{13}{22}$ 21) $4\frac{17}{21}$

Subtracting Mixed Numbers

1) $2\frac{8}{45}$ 8) $2\frac{22}{63}$ 15) $2\frac{8}{15}$ 22) $1\frac{13}{35}$

2) $4\frac{16}{35}$ 9) $1\frac{5}{24}$ 16) $1\frac{13}{34}$ 23) $1\frac{1}{24}$

3) $2\frac{4}{35}$ 10) $2\frac{11}{36}$ 17) $3\frac{11}{21}$ 24) $2\frac{11}{36}$

4) $6\frac{5}{8}$ 11) $4\frac{11}{18}$ 18) $1\frac{1}{3}$ 25) $2\frac{8}{25}$

5) $2\frac{11}{21}$ 12) $1\frac{13}{24}$ 19) $2\frac{1}{6}$ 26) $6\frac{5}{24}$

6) $1\frac{3}{4}$ 13) $1\frac{1}{8}$ 20) $2\frac{11}{18}$

7) $1\frac{31}{72}$ 14) $2\frac{5}{16}$ 21) $4\frac{13}{63}$

Multiplying Mixed Numbers

1) $1\frac{31}{32}$ 6) $23\frac{1}{3}$ 11) $23\frac{3}{4}$ 16) $14\frac{22}{35}$

2) $7\frac{11}{35}$ 7) $20\frac{7}{12}$ 12) $14\frac{13}{27}$ 17) $18\frac{5}{24}$

3) $2\frac{43}{72}$ 8) $22\frac{23}{40}$ 13) $18\frac{5}{12}$ 18) $16\frac{1}{4}$

4) $5\frac{7}{10}$ 9) $17\frac{11}{15}$ 14) $36\frac{2}{5}$ 19) $30\frac{2}{5}$

5) $8\frac{1}{2}$ 10) $12\frac{4}{21}$ 15) $17\frac{17}{24}$ 20) $11\frac{3}{8}$

21) $21\frac{8}{15}$ 23) $21\frac{7}{12}$ 25) $14\frac{9}{20}$

22) $13\frac{1}{7}$ 24) $25\frac{13}{35}$ 26) $26\frac{1}{15}$

Dividing Mixed Numbers

1) $3\frac{17}{26}$ 8) $3\frac{3}{80}$ 15) $2\frac{10}{77}$ 22) $1\frac{3}{14}$

2) $1\frac{39}{56}$ 9) $2\frac{53}{140}$ 16) $\frac{64}{95}$ 23) $2\frac{7}{24}$

3) $2\frac{33}{64}$ 10) $\frac{2}{3}$ 17) $1\frac{47}{68}$ 24) $1\frac{19}{56}$

4) $1\frac{49}{51}$ 11) $2\frac{88}{105}$ 18) $\frac{55}{64}$ 25) $\frac{84}{115}$

5) $1\frac{73}{75}$ 12) $\frac{95}{98}$ 19) $2\frac{13}{21}$ 26) $\frac{51}{77}$

6) $\frac{42}{55}$ 13) $1\frac{12}{31}$ 20) $1\frac{19}{32}$

7) $1\frac{77}{95}$ 14) $1\frac{53}{136}$ 21) $1\frac{31}{84}$

Chapter 2:

Decimals

Math Topics that you'll learn in this Chapter:

- ✓ Comparing Decimals
- ✓ Rounding Decimals
- ✓ Adding and Subtracting Decimals
- ✓ Multiplying and Dividing Decimals

Comparing Decimals

Compare. Use >, =, and <

1) 0.88 ☐ 0.088

2) 0.56 ☐ 0.57

3) 0.99 ☐ 0.89

4) 1.55 ☐ 1.65

5) 1.58 ☐ 1.75

6) 2.91 ☐ 2.85

7) 14.56 ☐ 1.456

8) 17.85 ☐ 17.89

9) 21.52 ☐ 21.052

10) 11.12 ☐ 11.03

11) 9.650 ☐ 9.65

12) 8.578 ☐ 8.568

13) 3.15 ☐ 0.315

14) 16.61 ☐ 16.16

15) 18.581 ☐ 8.991

16) 25.05 ☐ 2.505

17) 4.55 ☐ 4.65

18) 0.158 ☐ 1.58

19) 0.881 ☐ 0.871

20) 0.505 ☐ 0.510

21) 0.772 ☐ 0.777

22) 0.5 ☐ 0.500

23) 16.89 ☐ 15.89

24) 12.25 ☐ 12.35

25) 5.82 ☐ 5.69

26) 1.320 ☐ 1.032

27) 0.082 ☐ 0.088

28) 0.99 ☐ 0.099

29) 2.560 ☐ 1.950

30) 0.770 ☐ 0.707

31) 15.54 ☐ 1.554

32) 0.323 ☐ 0.332

Rounding Decimals

✎ *Round each number to the underlined place value.*

1) 2̲.814 =

2) 3.5̲62 =

3) 12.12̲5 =

4) 15̲.5 =

5) 1.98̲1 =

6) 14.2̲15 =

7) 17.54̲8 =

8) 25.50̲8 =

9) 31̲.089 =

10) 69.3̲45 =

11) 9.45̲7 =

12) 12̲.901 =

13) 2.65̲8 =

14) 32.5̲65 =

15) 6.05̲8 =

16) 98.10̲8 =

17) 27.7̲05 =

18) 36̲.75 =

19) 9.0̲8 =

20) 7.1̲85 =

21) 22.54̲7 =

22) 66.0̲98 =

23) 87̲.75 =

24) 18.5̲41 =

25) 10.25̲8 =

26) 13.4̲56 =

27) 71.08̲4 =

28) 29̲.23 =

29) 45.5̲5 =

30) 91̲.08 =

31) 83.4̲33 =

32) 74.6̲4 =

Adding and Subtracting Decimals

Solve.

1) $15.63 + 19.64 =$

2) $16.38 + 17.59 =$

3) $75.31 - 59.69 =$

4) $49.38 - 29.89 =$

5) $24.32 + 26.45 =$

6) $36.25 + 18.37 =$

7) $47.85 - 35.12 =$

8) $85.65 - 67.48 =$

9) $25.49 + 34.18 =$

10) $19.99 + 48.66 =$

11) $46.32 - 27.77 =$

12) $54.62 - 48.12 =$

13) $24.42 + 16.54 =$

14) $52.13 + 12.32 =$

15) $82.36 - 78.65 =$

16) $64.12 - 49.15 =$

17) $36.41 + 24.52 =$

18) $85.96 - 74.63 =$

19) $52.62 - 42.54 =$

20) $21.20 + 24.58 =$

21) $32.15 + 17.17 =$

22) $96.32 - 85.54 =$

23) $89.78 - 69.85 =$

24) $29.28 + 39.79 =$

25) $11.11 + 19.99 =$

26) $28.82 + 20.88 =$

27) $63.14 - 28.91 =$

28) $56.61 - 49.72 =$

29) $26.13 + 31.13 =$

30) $30.19 + 20.87 =$

31) $66.24 - 59.10 =$

32) $89.31 - 72.17 =$

Multiplying and Dividing Decimals

✎ *Solve.*

1) 11.2 × 0.4 =

2) 13.5 × 0.8 =

3) 42.2 ÷ 2 =

4) 54.6 ÷ 6 =

5) 23.1 × 0.3 =

6) 1.2 × 0.7 =

7) 5.5 ÷ 0.5 =

8) 64.8 ÷ 8 =

9) 1.4 × 0.5 =

10) 4.5 × 0.3 =

11) 88.8 ÷ 4 =

12) 10.5 ÷ 5 =

13) 2.2 × 0.3 =

14) 0.2 × 0.52 =

15) 95.7 ÷ 100 =

16) 36.6 ÷ 6 =

17) 3.2 × 2 =

18) 4.1 × 0.5 =

19) 68.4 ÷ 2 =

20) 27.9 ÷ 9 =

21) 3.5 × 4 =

22) 4.8 × 0.5 =

23) 6.4 ÷ 4 =

24) 72.8 ÷ 0.8 =

25) 1.8 × 3 =

26) 6.5 × 0.2 =

27) 93.6 ÷ 3 =

28) 45.15 ÷ 0.5 =

29) 13.2 × 0.4 =

30) 11.2 × 5 =

31) 7.2 ÷ 0.8 =

32) 96.4 ÷ 0.2 =

Answers – Chapter 2

Comparing Decimals

1) 0.88 > 0.088
2) 0.56 < 0.57
3) 0.99 > 0.89
4) 1.55 < 1.65
5) 1.58 < 1.75
6) 2.91 > 2.85
7) 14.56 > 1.456
8) 17.85 < 17.89
9) 21.52 > 21.052
10) 11.12 > 11.03
11) 9.650 = 9.65
12) 8.578 > 8.568
13) 3.15 > 0.315
14) 16.61 > 16.16
15) 18.581 > 8.991
16) 25.05 > 2.505
17) 4.55 < 4.65
18) 0.158 < 1.58
19) 0.881 > 0.871
20) 0.505 < 0.510
21) 0.772 < 0.777
22) 0.5 = 0.500
23) 16.89 > 15.89
24) 12.25 < 12.35
25) 5.82 > 5.69
26) 1.320 > 1.032
27) 0.082 < 0.088
28) 0.99 > 0.099
29) 2.560 > 1.950
30) 0.770 > 0.707
31) 15.54 > 1.554
32) 0.323 < 0.332

Rounding Decimals

1) 2.814 = 3
2) 3.562 = 3.56
3) 12.125 = 12.13
4) 15.5 = 16
5) 1.981 = 1.98
6) 14.215 = 14.2
7) 17.548 = 17.55
8) 25.508 = 25.51
9) 31.089 = 31
10) 69.345 = 69.3
11) 9.457 = 9.46
12) 12.901 = 13
13) 2.658 = 2.66
14) 32.565 = 32.6
15) 6.058 = 6.06
16) 98.108 = 98.11
17) 27.705 = 27.7
18) 36.75 = 37
19) 9.08 = 9.1
20) 7.185 = 7.2
21) 22.547 = 22.55
22) 66.098 = 66.1
23) 87.75 = 88
24) 18.541 = 18.5
25) 10.258 = 10.26
26) 13.456 = 13.5
27) 71.084 = 71.08
28) 29.23 = 29
29) 45.55 = 45.6
30) 91.08 = 91
33) 83.433 = 83
34) 74.64 = 74.6

Adding and Subtracting Decimals

1) 35.27
2) 33.97
3) 15.62
4) 19.49
5) 50.77
6) 54.62
7) 12.73
8) 18.17
9) 59.67
10) 68.65
11) 18.55
12) 6.5
13) 40.96
14) 64.45
15) 3.71
16) 14.97
17) 60.93
18) 11.33
19) 10.08
20) 45.78
21) 49.32
22) 10.78
23) 19.93
24) 69.07
25) 31.1
26) 49.7
27) 34.23
28) 6.89
29) 57.26
30) 51.06

31) 7.14   32) 17.14

Multiplying and Dividing Decimals

1) 4.48
2) 10.8
3) 21.1
4) 9.1
5) 6.93
6) 0.84
7) 1.1
8) 8.1
9) 0.7
10) 1.35
11) 22.2
12) 2.1
13) 0.66
14) 0.104
15) 0.957
16) 6.1
17) 6.4
18) 2.05
19) 34.2
20) 3.1
21) 14
22) 2.4
23) 1.6
24) 91
25) 5.4
26) 1.3
27) 31.2
28) 90.3
29) 5.28
30) 56
31) 9
32) 482

Chapter 3:
Integers and Order of Operations

Math Topics that you'll learn in this Chapter:

- ✓ Adding and Subtracting Integers
- ✓ Multiplying and Dividing Integers
- ✓ Order of Operations
- ✓ Integers and Absolute Value

Adding and Subtracting Integers

✎ *Solve.*

1) $-(8) + 13 =$

2) $17 - (-12 - 8) =$

3) $(-15) + (-4) =$

4) $(-14) + (-8) + 9 =$

5) $-(23) + 19 =$

6) $(-7 + 5) - 9 =$

7) $28 + (-32) =$

8) $(-11) + (-9) + 5 =$

9) $25 - (8 - 7) =$

10) $-(29) + 17 =$

11) $(-38) + (-3) + 29 =$

12) $15 - (-7 + 9) =$

13) $24 - (8 - 2) =$

14) $(-7 + 4) - 9 =$

15) $(-17) + (-3) + 9 =$

16) $(-26) + (-7) + 8 =$

17) $(-9) + (-11) =$

18) $8 - (-23 - 13) =$

19) $(-16) + (-2) =$

20) $25 - (7 - 4) =$

21) $23 + (-12) =$

22) $(-18) + (-6) =$

23) $17 - (-21 - 7) =$

24) $-(28) - (-16) + 5 =$

25) $(-9 + 4) - 8 =$

26) $(-28) + (-6) + 17 =$

27) $-(21) - (-15) + 9 =$

28) $(-31) + (-6) =$

29) $(-17) + (-11) + 14 =$

30) $(-29) + (-10) + 13 =$

31) $-(24) - (-12) + 5 =$

32) $8 - (-19 - 10) =$

Multiplying and Dividing Integers

✏️ **Solve.**

1) $(-9) \times (-8) =$

2) $6 \times (-6) =$

3) $49 \div (-7) =$

4) $(-64) \div 8 =$

5) $(4) \times (-6) =$

6) $(-9) \times (-11) =$

7) $(10) \div (-5) =$

8) $144 \div (-12) =$

9) $(10) \times (-2) =$

10) $(-8) \times (-2) \times 5 =$

11) $(8) \div (-2) =$

12) $45 \div (-15) =$

13) $(5) \times (-7) =$

14) $(-6) \times (-5) \times 4 =$

15) $(12) \div (-6) =$

16) $(14) \div (-7) =$

17) $196 \div (-14) =$

18) $(27 - 13) \times (-2) =$

19) $125 \div (-5) =$

20) $66 \div (-6) =$

21) $(-6) \times (-5) \times 3 =$

22) $(15 - 6) \times (-3) =$

23) $(32 - 24) \div (-4) =$

24) $72 \div (-6) =$

25) $(-14 + 8) \times (-7) =$

26) $(-3) \times (-9) \times 3 =$

27) $84 \div (-12) =$

28) $(-12) \times (-10) =$

29) $25 \times (-4) =$

30) $(-3) \times (-5) \times 5 =$

31) $(15) \div (-3) =$

32) $(-18) \div (3) =$

Order of Operation

✎ **Calculate.**

1) $18 + (32 \div 4) =$

2) $(3 \times 8) \div (-2) =$

3) $67 - (4 \times 8) =$

4) $(-11) \times (8 - 3) =$

5) $(18 - 7) \times (6) =$

6) $(6 \times 10) \div (12 + 3) =$

7) $(13 \times 2) - (24 \div 6) =$

8) $(-5) + (4 \times 3) + 8 =$

9) $(4 \times 2^3) + (16 - 9) =$

10) $(3^2 \times 7) \div (-2 + 1) =$

11) $[-2(48 \div 2^3)] - 6 =$

12) $(-4) + (7 \times 8) + 18 =$

13) $(3 \times 7) + (16 - 7) =$

14) $[3^3 \times (48 \div 2^3)] \div (-2) =$

15) $(14 \times 3) - (3^4 \div 9) =$

16) $(96 \div 12) \times (-3) =$

17) $(48 \div 2^2) \times (-2) =$

18) $(56 \div 7) \times (-5) =$

19) $(-2^2) + (7 \times 9) - 21 =$

20) $(2^4 - 9) \times (-6) =$

21) $[4^3 \times (50 \div 5^2)] \div (-16) =$

22) $(3^2 \times 4^2) \div (-4 + 2) =$

23) $6^2 - (-6 \times 4) + 3 =$

24) $4^2 - (5^2 \times 3) =$

25) $(-4) + (12^2 \div 3^2) - 7^2 =$

26) $(3^2 \times 5) + (-5^2 - 9) =$

27) $2[(3^2 \times 5) \times (-6)] =$

28) $(11^2 - 2^2) - (-7^2) =$

29) $(2^3 \times 3) - (49 \div 7) =$

30) $3[(3^2 \times 5) + (25 \div 5)] =$

31) $(6^2 \times 5) \div (-5) =$

32) $2^2[(6^3 \div 12) - (3^4 \div 27)] =$

Integers and Absolute Value

✍ *Calculate.*

1) $5 - |8 - 12| =$

2) $|15| - \frac{|-1|}{4} =$

3) $\frac{|9 \times -6|}{18} \times \frac{|-2|}{8} =$

4) $|13 \times 3| + \frac{|-72|}{9} =$

5) $4 - |11 - 18| - |3| =$

6) $|18| - \frac{|-1|}{4} =$

7) $\frac{|5 \times -8|}{10} \times \frac{|-22|}{11} =$

8) $|9 \times 3| + \frac{|-36|}{4} =$

9) $|-42 + 7| \times \frac{|-2 \times 5|}{10} =$

10) $6 - |17 - 11| - |5| =$

11) $|13| - \frac{|-5|}{6} =$

12) $\frac{|9 \times -4|}{12} \times \frac{|-45|}{9} =$

13) $|-75 + 50| \times \frac{|-4 \times 5|}{5} =$

14) $\frac{|-26|}{13} \times \frac{|-32|}{8} =$

15) $14 - |8 - 18| - |-12| =$

16) $|29| - \frac{|-20|}{5} =$

17) $\frac{|3 \times 8|}{2} \times \frac{|-33|}{3} =$

18) $|-45 + 15| \times \frac{|-12 \times 5|}{6} =$

19) $\frac{|-50|}{5} \times \frac{|-77|}{11} =$

20) $12 - |2 - 7| - |15| =$

21) $|18| - \frac{|-45|}{15} =$

22) $\frac{|7 \times 8|}{4} \times \frac{|-48|}{12} =$

23) $\frac{|30 \times 2|}{3} \times |-12| =$

24) $\frac{|-36|}{9} \times \frac{|-80|}{8} =$

25) $|-35 + 8| \times \frac{|-9 \times 5|}{15} =$

26) $|19| - \frac{|-18|}{2} =$

27) $14 - |11 - 23| + |2| =$

28) $|-39 + 7| \times \frac{|-4 \times 6|}{3} =$

Answers – Chapter 3

Adding and Subtracting Integers

1) 5
2) 37
3) −19
4) −13
5) −4
6) −11
7) −4
8) −15
9) 24
10) −12
11) −12
12) 13
13) 18
14) −12
15) −11
16) −25
17) −20
18) 44
19) −18
20) 22
21) 11
22) −24
23) 45
24) −7
25) −13
26) −17
27) 3
28) −37
29) −14
30) −26
31) −7
32) 37

Multiplying and Dividing Integers

1) 72
2) −36
3) −7
4) −8
5) −24
6) 99
7) −2
8) −12
9) −20
10) 80
11) −4
12) −3
13) −35
14) 150
15) −2
16) −2
17) −14
18) −28
19) 25
20) −11
21) 90
22) −27
23) −2
24) −12
25) 42
26) 81
27) −7
28) 120
29) −100
30) 75
31) −5
32) −6

Order of Operation

1) 26
2) −12
3) 35
4) −55
5) 66
6) 4
7) 22
8) 15
9) 39
10) −63
11) −18
12) 70
13) 30
14) −81
15) 33
16) −24
17) −24
18) −40
19) 38
20) −42
21) −8
22) −72
23) 63
24) −59
25) −37
26) 11
27) −540
28) 166
29) 17
30) 150

31) −36   32) 60

Integers and Absolute Value

1) 1
2) 11
3) 9
4) 47
5) −6
6) 15
7) 8
8) 36
9) 35
10) −5
11) 4
12) 15
13) 100
14) 8
15) −8
16) 25
17) 132
18) 300
19) 70
20) −8
21) 15
22) 56
23) 240
24) 40
25) 81
26) 10
27) 4
28) 256

Chapter 4:

Ratios and Proportions

Math Topics that you'll learn in this Chapter:

- ✓ Simplifying Ratios
- ✓ Proportional Ratios
- ✓ Similarity and Ratios
- ✓ Simple Interest

Simplifying Ratios

✏️ *Simplify each ratio.*

1) $3:27 = \underline{}:\underline{}$

2) $2:8 = \underline{}:\underline{}$

3) $\dfrac{4}{28} = -$

4) $\dfrac{16}{40} = -$

5) $10:30 = \underline{}:\underline{}$

6) $5:30 = \underline{}:\underline{}$

7) $\dfrac{34}{38} = -$

8) $\dfrac{45}{63} = -$

9) $10:45 = \underline{}:\underline{}$

10) $20:30 = \underline{}:\underline{}$

11) $\dfrac{40}{64} = -$

12) $\dfrac{10}{110} = -$

13) $8:12 = \underline{}:\underline{}$

14) $16:20 = \underline{}:\underline{}$

15) $\dfrac{24}{48} = -$

16) $\dfrac{21}{77} = -$

17) $8:24 = \underline{}:\underline{}$

18) $9 \text{ to } 36 = \underline{}:\underline{}$

19) $\dfrac{64}{72} = -$

20) $\dfrac{45}{60} = -$

21) $12:15 = \underline{}:\underline{}$

22) $18:54 = \underline{}:\underline{}$

23) $\dfrac{36}{54} = -$

24) $\dfrac{48}{104} = -$

25) $15:75 = \underline{}:\underline{}$

26) $16:48 = \underline{}:\underline{}$

27) $\dfrac{15}{65} = -$

28) $\dfrac{44}{52} = -$

www.EffortlessMath.com

Proportional Ratios

Solve each proportion for x.

1) $\dfrac{4}{7} = \dfrac{16}{x}$, $x =$ _____

2) $\dfrac{4}{9} = \dfrac{x}{18}$, $x =$ _____

3) $\dfrac{3}{5} = \dfrac{24}{x}$, $x =$ _____

4) $\dfrac{3}{10} = \dfrac{x}{50}$, $x =$ _____

5) $\dfrac{3}{11} = \dfrac{15}{x}$, $x =$ _____

6) $\dfrac{6}{15} = \dfrac{x}{45}$, $x =$ _____

7) $\dfrac{6}{19} = \dfrac{12}{x}$, $x =$ _____

8) $\dfrac{7}{16} = \dfrac{x}{32}$, $x =$ _____

9) $\dfrac{18}{21} = \dfrac{54}{x}$, $x =$ _____

10) $\dfrac{13}{15} = \dfrac{39}{x}$, $x =$ _____

11) $\dfrac{9}{13} = \dfrac{72}{x}$, $x =$ _____

12) $\dfrac{8}{30} = \dfrac{x}{180}$, $x =$ _____

13) $\dfrac{3}{19} = \dfrac{9}{x}$, $x =$ _____

14) $\dfrac{1}{3} = \dfrac{x}{90}$, $x =$ _____

15) $\dfrac{25}{45} = \dfrac{x}{9}$, $x =$ _____

16) $\dfrac{1}{6} = \dfrac{9}{x}$, $x =$ _____

17) $\dfrac{7}{9} = \dfrac{63}{x}$, $x =$ _____

18) $\dfrac{54}{72} = \dfrac{x}{8}$, $x =$ _____

19) $\dfrac{32}{40} = \dfrac{4}{x}$, $x =$ _____

20) $\dfrac{21}{42} = \dfrac{x}{6}$, $x =$ _____

21) $\dfrac{56}{72} = \dfrac{7}{x}$, $x =$ _____

22) $\dfrac{1}{14} = \dfrac{x}{42}$, $x =$ _____

23) $\dfrac{5}{7} = \dfrac{75}{x}$, $x =$ _____

24) $\dfrac{30}{48} = \dfrac{x}{8}$, $x =$ _____

25) $\dfrac{36}{88} = \dfrac{9}{x}$, $x =$ _____

26) $\dfrac{62}{68} = \dfrac{x}{34}$, $x =$ _____

27) $\dfrac{42}{60} = \dfrac{x}{10}$, $x =$ _____

28) $\dfrac{8}{9} = \dfrac{x}{108}$, $x =$ _____

29) $\dfrac{46}{69} = \dfrac{x}{3}$, $x =$ _____

30) $\dfrac{99}{121} = \dfrac{x}{11}$, $x =$ _____

31) $\dfrac{19}{21} = \dfrac{x}{63}$, $x =$ _____

32) $\dfrac{11}{12} = \dfrac{x}{48}$, $x =$ _____

Create Proportion

✎ *State if each pair of ratios form a proportion.*

1) $\frac{5}{8}$ and $\frac{25}{50}$

2) $\frac{2}{11}$ and $\frac{4}{22}$

3) $\frac{2}{5}$ and $\frac{8}{20}$

4) $\frac{3}{11}$ and $\frac{9}{33}$

5) $\frac{5}{10}$ and $\frac{15}{30}$

6) $\frac{4}{13}$ and $\frac{8}{24}$

7) $\frac{6}{9}$ and $\frac{24}{36}$

8) $\frac{7}{12}$ and $\frac{14}{20}$

9) $\frac{3}{8}$ and $\frac{27}{72}$

10) $\frac{12}{20}$ and $\frac{36}{60}$

11) $\frac{11}{12}$ and $\frac{55}{60}$

12) $\frac{12}{15}$ and $\frac{24}{25}$

13) $\frac{15}{19}$ and $\frac{20}{38}$

14) $\frac{10}{14}$ and $\frac{40}{56}$

15) $\frac{11}{13}$ and $\frac{44}{39}$

16) $\frac{15}{16}$ and $\frac{30}{32}$

17) $\frac{17}{19}$ and $\frac{34}{48}$

18) $\frac{5}{18}$ and $\frac{15}{54}$

19) $\frac{3}{14}$ and $\frac{18}{42}$

20) $\frac{7}{11}$ and $\frac{14}{32}$

21) $\frac{8}{11}$ and $\frac{32}{44}$

22) $\frac{9}{13}$ and $\frac{18}{26}$

✎ *Solve.*

23) The ratio of boys to girls in a class is 5:6. If there are 25 boys in the class, how many girls are in that class? _____

24) The ratio of red marbles to blue marbles in a bag is 4:7. If there are 77 marbles in the bag, how many of the marbles are red? _____

25) You can buy 8 cans of green beans at a supermarket for $3.20. How much does it cost to buy 48 cans of green beans? _____

Similarity and Ratios

Each pair of figures is similar. Find the missing side.

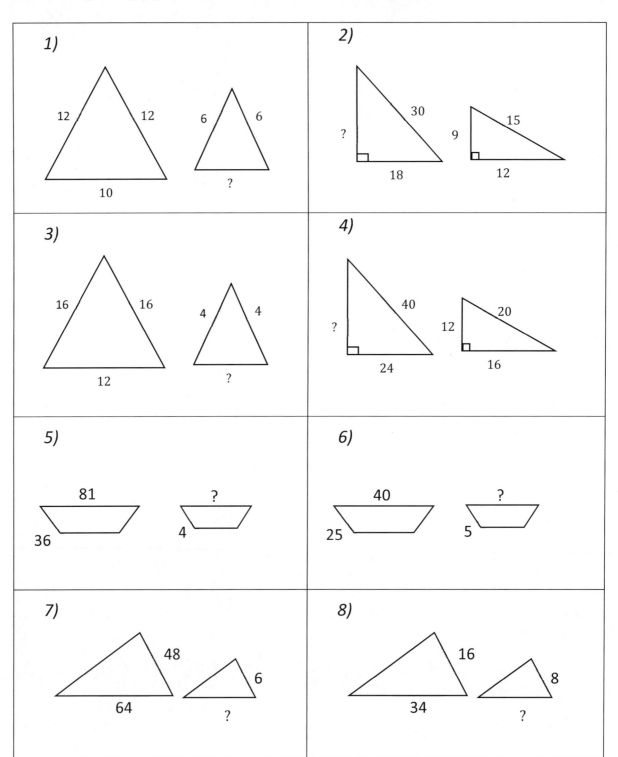

Simple Interest

Determine the simple interest for these loans.

1) $440 at 5% for 6 years. $___
2) $460 at 2.5% for 4 years. $___
3) $500 at 3% for 5 years. $___
4) $550 at 9% for 2 years. $___
5) $690 at 5% for 6 months. $___
6) $620 at 7% for 3 years. $___
7) $650 at 4.5% for 10 years. $___
8) $850 at 4% for 2 years. $___
9) $640 at 7% for 3 years. $___
10) $300 at 9% for 9 months. $___
11) $760 at 8% for 2 years. $___
12) $910 at 5% for 5 years. $___
13) $540 at 3% for 6 years. $___
14) $780 at 2.5% for 4 years. $___
15) $1,600 at 7% for 3 months. $___
16) $310 at 4% for 4 years. $___
17) $950 at 6% for 5 years. $___
18) $280 at 8% for 7 years. $___
19) $310 at 6% for 3 years. $___
20) $990 at 5% for 4 months. $___
21) $380 at 6% for 5 years. $___

22) $580 at 6% for 4 years. $___
23) $1,200 at 4% for 5 years. $___
24) $3,100 at 5% for 6 years. $___
25) $5,200 at 8% for 2 years. $___
26) $1,400 at 4% for 3 years. $___
27) $300 at 3% for 8 months. $___
28) $150 at 3.5% for 4 years. $___
29) $170 at 6% for 2 years. $___
30) $940 at 8% for 5 years. $___
31) $960 at 1.5% for 8 years. $___
32) $240 at 5% for 4 months. $___
33) $280 at 2% for 5 years. $___
34) $880 at 3% for 2 years. $___
35) $2,200 at 4.5% for 2 years. $___
36) $2,400 at 7% for 3 years. $___
37) $1,800 at 5% for 6 months. $___
38) $190 at 4% for 2 years. $___
39) $560 at 7% for 4 years. $___
40) $720 at 8% for 2 years. $___
41) $780 at 5% for 8 years. $___
42) $880 at 6% for 3 months. $___

www.EffortlessMath.com

Answers – Chapter 4

Simplifying Ratios

1) 1 : 9
2) 1 : 4
3) $\frac{1}{7}$
4) $\frac{2}{5}$
5) 1 : 3
6) 1 : 6
7) $\frac{17}{19}$
8) $\frac{5}{7}$
9) 2 : 9
10) 2 : 3
11) $\frac{5}{8}$
12) $\frac{1}{11}$
13) 2 : 3
14) 4 : 5
15) $\frac{1}{2}$
16) $\frac{3}{11}$
17) 1 : 6
18) 1 to 4
19) $\frac{8}{9}$
20) $\frac{3}{4}$
21) 4 : 5
22) 1 : 3
23) $\frac{2}{3}$
24) $\frac{6}{13}$
25) 1 : 5
26) 1 : 3
27) $\frac{3}{13}$
28) $\frac{11}{13}$

Proportional Ratios

1) $x = 28$
2) $x = 8$
3) $x = 40$
4) $x = 15$
5) $x = 55$
6) $x = 18$
7) $x = 38$
8) $x = 14$
9) $x = 63$
10) $x = 45$
11) $x = 104$
12) $x = 48$
13) $x = 57$
14) $x = 30$
15) $x = 5$
16) $x = 54$
17) $x = 81$
18) $x = 6$
19) $x = 5$
20) $x = 3$
21) $x = 9$
22) $x = 3$
23) $x = 105$
24) $x = 5$
25) $x = 22$
26) $x = 31$
27) $x = 7$
28) $x = 96$
29) $x = 2$
30) $x = 9$
31) $x = 57$
32) $x = 44$

Create Proportion

1) No
2) Yes
3) Yes
4) Yes
5) Yes
6) No
7) Yes
8) No
9) Yes
10) Yes
11) Yes
12) No
13) No
14) Yes
15) No
16) Yes
17) Yes
18) Yes
19) No
20) No
21) Yes

22) Yes
23) 30 *girls*
24) 28 *red marbles*
25) $19.20

Similarity and Ratios

1) 5
2) 24
3) 3
4) 32
5) 9
6) 8
7) 8
8) 17

Simple Interest

1) $132
2) $46
3) $75
4) $99
5) $17.25
6) $130.20
7) $292.50
8) $68
9) $134.40
10) $20.25
11) $121.60
12) $227.50
13) $97.20
14) $78
15) $28
16) $49.60
17) $285
18) $156.80
19) $55.80
20) $198
21) $114
22) $139.20
23) $240
24) $930
25) $832
26) $168
27) $6
28) $21
29) $20.40
30) $376
31) $115.20
32) $4
33) $28
34) $52.80
35) $198
36) $504
37) $45
38) $15.20
39) $156.80
40) $115.20
41) $312
42) $13.20

Chapter 5:

Percentage

Math Topics that you'll learn in this Chapter:

✓ Percent Problems

✓ Percent of Increase and Decrease

✓ Discount, Tax and Tip

Percent Problems

Solve each problem.

1) What is 5 percent of 300? ____
2) What is 15 percent of 600? ____
3) What is 12 percent of 450? ____
4) What is 30 percent of 240? ____
5) What is 60 percent of 850? ____
6) 63 is what percent of 300? ____%
7) 80 is what percent of 400? ____%
8) 70 is what percent of 700? ____%
9) 84 is what percent of 600? ____%
10) 90 is what percent of 300? ____%
11) 24 is what percent of 150? ____%
12) 12 is what percent of 80? ____%
13) 4 is what percent of 50? ____%
14) 110 is what percent of 500? ____%
15) 16 is what percent of 400? ____%
16) 39 is what percent of 300? ____%
17) 56 is what percent of 200? ____%
18) 30 is what percent of 500? ____%
19) 84 is what percent of 700? ____%
20) 40 is what percent of 500? ____%
21) 26 is what percent of 100? ____%
22) 45 is what percent of 900? ____%
23) 60 is what percent of 400? ____%
24) 18 is what percent of 900? ____%
25) 75 is what percent of 250? ____%
26) 27 is what percent of 900? ____%
27) 49 is what percent of 700? ____%
28) 81 is what percent of 900? ____%
29) 90 is what percent of 500? ____%
30) 82 is 20 percent of what number? ____
31) 14 is 35 percent of what number? ____
32) 90 is 6 percent of what number? ____
33) 80 is 40 percent of what number? ____
34) 90 is 15 percent of what number? ____
35) 28 is 7 percent of what number? ____
36) 54 is 18 percent of what number? ____
37) 72 is 24 percent of what number? ____

Percent of Increase and Decrease

Solve each percent of change word problem.

1) Bob got a raise, and his hourly wage increased from $24 to $36. What is the percent increase? _____ %

2) The price of gasoline rose from $2.20 to $2.42 in one month. By what percent did the gas price rise? _____ %

3) In a class, the number of students has been increased from 30 to 39. What is the percent increase? _____ %

4) The price of a pair of shoes increases from $28 to $35. What is the percent increase? ___ %

5) In a class, the number of students has been decreased from 24 to 18. What is the percentage decrease? _____ %

6) Nick got a raise, and his hourly wage increased from $50 to $55. What is the percent increase? _____ %

7) A coat was originally priced at $80. It went on sale for $70.40. What was the percent that the coat was discounted? _____ %

8) The price of a pair of shoes increases from $8 to $12. What is the percent increase? ___ %

9) A house was purchased in 2002 for $180,000. It is now valued at $144,000. What is the rate (percent) of depreciation for the house? _____ %

10) The price of gasoline rose from $3.00 to $3.15 in one month. By what percent did the gas price rise? _____ %

Discount, Tax and Tip

✍ Find the missing values.

1) Original price of a computer: $400

   Tax: 5%, Selling price: $_____

2) Original price of a sofa: $600

   Tax: 12%, Selling price: $_____

3) Original price of a table: $550

   Tax: 18%, Selling price: $_____

4) Original price of a cell phone: $700

   Tax: 20%, Selling price: $_____

5) Original price of a printer: $400

   Tax: 22%, Selling price: $_____

6) Original price of a computer: $600

   Tax: 15%, Selling price: $_____

7) Restaurant bill: $24.00

   Tip: 25%, Final amount: $_____

8) Original price of a cell phone: $300

   Tax: 8%, Selling price: $_____

9) Original price of a carpet: $800

   Tax: 25%, Selling price: $_____

10) Original price of a camera: $200

    Discount: 35%, Selling price: $_____

11) Original price of a dress: $500

    Discount: 10%, Selling price: $_____

12) Original price of a monitor: $400

    Discount: 5%, Selling price: $_____

13) Original price of a laptop: $900

    Discount: 20%, Selling price: $_____

14) Restaurant bill: $54.00

    Tip: 20%, Final amount: $_____

www.EffortlessMath.com

Answers – Chapter 5

Percent Problems

1) 15
2) 90
3) 54
4) 72
5) 510
6) 21%
7) 20%
8) 10%
9) 14%
10) 30%
11) 16%
12) 15%
13) 8%
14) 22%
15) 4%
16) 13%
17) 28%
18) 6%
19) 12%
20) 8%
21) 26%
22) 5%
23) 15%
24) 2%
25) 30%
26) 3%
27) 7%
28) 9%
29) 18%
30) 410
31) 40
32) 1,500
33) 200
34) 600
35) 400
36) 300
37) 300

Percent of Increase and Decrease - Answers

1) 50%
2) 10%
3) 30%
4) 25%
5) 25%
6) 10%
7) 12%
8) 50%
9) 20%
10) 5%

Discount, Tax and Tip - Answers

1) $420
2) $672
3) $649
4) $840
5) $488
6) $690
7) $30.00
8) $324
9) $1,000
10) $130
11) $450
12) $380
13) $720
14) $64.80

Chapter 6:

Expressions and Variables

Math Topics that you'll learn in this Chapter:

- ✓ Simplifying Variable Expressions
- ✓ Simplifying Polynomial Expressions
- ✓ Evaluating One Variable
- ✓ Evaluating Two Variables
- ✓ The Distributive Property

Simplifying Variable Expressions

✎ *Simplify and write the answer.*

1) $3x + 5 + 2x =$

2) $7x + 3 - 3x =$

3) $-2 - x^2 - 6x^2 =$

4) $(-6)(8x - 4) =$

5) $3 + 10x^2 + 2x =$

6) $8x^2 + 6x + 7x^2 =$

7) $2x^2 - 5x - 7x =$

8) $x - 3 + 5 - 3x =$

9) $2 - 3x + 12 - 2x =$

10) $5x^2 - 12x^2 + 8x =$

11) $2x^2 + 6x + 3x^2 =$

12) $2x^2 - 2x - x =$

13) $2x^2 - (-8x + 6) = 2$

14) $4x + 6(2 - 5x) =$

15) $10x + 8(10x - 6) =$

16) $9(-2x - 6) - 5 =$

17) $32x - 4 + 23 + 2x =$

18) $8x - 12x - x^2 + 13 =$

19) $(-6)(8x - 4) + 10x =$

20) $14x - 5(5 - 8x) =$

21) $23x + 4(9x + 3) + 12 =$

22) $3(-7x + 5) + 20x =$

23) $12x - 3x(x + 9) =$

24) $7x + 5x(3 - 3x) =$

25) $5x(-8x + 12) + 14x =$

26) $40x + 12 + 2x^2 =$

27) $5x(x - 3) - 10 =$

28) $8x - 7 + 8x + 2x^2 =$

29) $7x - 3x^2 - 5x^2 - 3 =$

30) $4 + x^2 - 6x^2 - 12x =$

31) $12x + 8x^2 + 2x + 20 =$

32) $23 + 15x^2 + 8x - 4x^2 =$

Simplifying Polynomial Expressions

✎ *Simplify and write the answer.*

1) $(2x^3 + 5x^2) - (12x + 2x^2) =$ _____

2) $(-x^5 + 2x^3) - (3x^3 + 6x^2) =$ _____

3) $(12x^4 + 4x^2) - (2x^2 - 6x^4) =$ _____

4) $4x - 3x^2 - 2(6x^2 + 6x^3) =$ _____

5) $(2x^3 - 3) + 3(2x^2 - 3x^3) =$ _____

6) $4(4x^3 - 2x) - (3x^3 - 2x^4) =$ _____

7) $2(4x - 3x^3) - 3(3x^3 + 4x^2) =$ _____

8) $(2x^2 - 2x) - (2x^3 + 5x^2) =$ _____

9) $2x^3 - (4x^4 + 2x) + x^2 =$ _____

10) $x^4 - 9(x^2 + x) - 5x =$ _____

11) $(-2x^2 - x^4) + (4x^4 - x^2) =$ _____

12) $4x^2 - 5x^3 + 15x^4 - 12x^3 =$ _____

13) $2x^2 - 5x^4 + 14x^4 - 11x^3 =$ _____

14) $2x^2 + 5x^3 - 7x^2 + 12x =$ _____

15) $2x^4 - 5x^5 + 8x^4 - 8x^2 =$ _____

16) $5x^3 + 17x - 5x^2 - 2x^3 =$ _____

www.EffortlessMath.com

Evaluating One Variable

✎ *Evaluate each expression using the value given.*

1) $x = 3 \Rightarrow 6x - 9 =$

2) $x = 2 \Rightarrow 7x - 10 =$

3) $x = 1 \Rightarrow 5x + 2 =$

4) $x = 2 \Rightarrow 3x + 9 =$

5) $x = 4 \Rightarrow 4x - 8 =$

6) $x = 2 \Rightarrow 5x - 2x + 10 =$

7) $x = 3 \Rightarrow 2x - x - 6 =$

8) $x = 4 \Rightarrow 6x - 3x + 4 =$

9) $x = -2 \Rightarrow 4x - 6x - 5 =$

10) $x = -1 \Rightarrow 3x - 5x + 11 =$

11) $x = 1 \Rightarrow x - 7x + 12 =$

12) $x = 2 \Rightarrow 2(-3x + 4) =$

13) $x = 3 \Rightarrow 4(-5x - 2) =$

14) $x = 2 \Rightarrow 5(-2x - 4) =$

15) $x = -2 \Rightarrow 3(-4x - 5) =$

16) $x = 3 \Rightarrow 8x + 5 =$

17) $x = -3 \Rightarrow 12x + 9 =$

18) $x = -1 \Rightarrow 9x - 8 =$

19) $x = 2 \Rightarrow 16x - 10 =$

20) $x = 1 \Rightarrow 4x + 3 =$

21) $x = 5 \Rightarrow 7x - 2 =$

22) $x = 7 \Rightarrow 28 - x =$

23) $x = 3 \Rightarrow 5x - 10 =$

24) $x = 12 \Rightarrow 40 - 2x =$

25) $x = 2 \Rightarrow 11x - 2 =$

26) $x = 3 \Rightarrow 2x - x + 10 =$

Evaluating Two Variables

✏️ *Evaluate each expression using the values given.*

1) $2x + 3y, x = 2, y = 3$

2) $3x + 4y, x = -1, y = -2$

3) $x + 6y, x = 3, y = 1$

4) $2a - (15 - b), a = 2, b = 3$

5) $4a - (6 - 3b), a = 1, b = 4$

6) $a - (8 - 2b), a = 2, b = 5$

7) $3z + 21 + 5k, z = 4, k = 1$

8) $-7a + 4b, a = 6, b = 3$

9) $-4a + 3b, a = 2, b = 4$

10) $-6a + 6b, a = 4, b = 3$

11) $-8a + 2b, a = 4, b = 6$

12) $4x + 6y, x = 6, y = 3$

13) $2x + 9y, x = 8, y = 1$

14) $x - 7y, x = 9, y = 4$

15) $5x - 4y, x = 6, y = 3$

16) $2z + 14 + 8k, z = 4, k = 1$

17) $6x + 3y, x = 3, y = 8$

18) $5a - 6b, a = -3, b = -1$

19) $8a + 4b, a = -4, b = 3$

20) $-2a - b, a = 4, b = 9$

21) $-7a + 3b, a = 4, b = 3$

22) $-5a + 9b, a = 7, b = 1$

The Distributive Property

✍ *Use the distributive property to simply each expression.*

1) $(-3)(12x + 3) =$

2) $(-4x + 5)(-6) =$

3) $13(-4x + 2) =$

4) $7(6 - 3x) =$

5) $(6 - 5x)(-4) =$

6) $9(8 - 2x) =$

7) $(-4x + 6)5 =$

8) $(-2x + 7)(-8) =$

9) $8(-4x + 7) =$

10) $(-9x + 5)(-3) =$

11) $8(-x + 9) =$

12) $7(2 - 6x) =$

13) $(-12x + 4)(-3) =$

14) $(-6)(-10x + 6) =$

15) $(-5)(5 - 11x) =$

16) $9(4 - 8x) =$

17) $(-6x + 2)7 =$

18) $(-9)(1 - 12x) =$

19) $(-3)(4 - 6x) =$

20) $(2 - 8x)(-2) =$

21) $20(2 - x) =$

22) $12(-4x + 3) =$

23) $15(2 - 3x) =$

24) $(-4x + 5)2 =$

25) $(-11x + 8)(-2) =$

26) $14(5 - 8x) =$

Answers – Chapter 6

Simplifying Variable Expressions

1) $5x + 5$
2) $4x + 3$
3) $-7x^2 - 2$
4) $-48x + 24$
5) $10x^2 + 2x + 3$
6) $15x^2 + 6x$
7) $2x^2 - 12x$
8) $-2x + 2$
9) $-5x + 14$
10) $-7x^2 + 8x$
11) $5x^2 + 6x$
12) $2x^2 - 3x$
13) $2x^2 + 8x - 6$
14) $-26x + 12$
15) $90x - 48$
16) $-18x - 59$
17) $34x + 19$
18) $-x^2 - 4x + 13$
19) $-38x + 24$
20) $54x - 25$
21) $59x + 24$
22) $-x + 15$
23) $-3x^2 - 15x$
24) $-15x^2 + 22x$
25) $-40x^2 + 74x$
26) $2x^2 + 40x + 12$
27) $5x^2 - 15x - 10$
28) $2x^2 + 16x - 7$
29) $-8x^2 + 7x - 3$
30) $-5x^2 - 12x + 4$
31) $8x^2 + 14x + 20$
32) $11x^2 + 8x + 23$

Simplifying Polynomial Expressions

1) $2x^3 + 3x^2 - 12x$
2) $-x^5 - x^3 - 6x^2$
3) $18x^4 + 2x^2$
4) $-12x^3 - 15x^2 + 4x$
5) $-7x^3 + 6x^2 - 3$
6) $2x^4 + 13x^3 - 8x$
7) $-15x^3 - 12x^2 + 8x$
14) $5x^3 - 5x^2 + 12x$
15) $-5x^5 + 10x^4 - 8x^2$
8) $-2x^3 - 3x^2 - 2x$
9) $-4x^4 + 2x^3 + x^2 - 2x$
10) $x^4 - 9x^2 - 14x$
11) $3x^4 - 3x^2$
12) $15x^4 - 17x^3 + 4x^2$
13) $9x^4 - 11x^3 + 2x^2$
16) $3x^3 - 5x^2 + 17x$

Evaluating One Variable

1) 9
2) 4
3) 7
4) 15
5) 8
6) 16
7) -3
8) 16
9) -1
10) 13
11) 6
12) -4
13) -68
14) -40
15) 9
16) 29
17) -27
18) -17
19) 22
20) 7
21) 33

22) 21
23) 5
24) 16
25) 20
26) 13

Evaluating Two Variables

1) 13
2) −11
3) 9
4) −8
5) 10
6) 4
7) 38
8) −30
9) 4
10) 6
11) −20
12) 42
13) 25
14) −19
15) 18
16) 30
17) 42
18) −9
19) −20
20) −17
21) −19
22) −26

The Distributive Property

1) $-36x - 9$
2) $24x - 30$
3) $-52x + 26$
4) $-21x + 42$
5) $20x - 24$
6) $-18x + 72$
7) $-20x + 30$
8) $16x - 56$
9) $-32x + 56$
10) $27x - 15$
11) $-8x + 72$
12) $-42x + 14$
13) $36x - 12$
14) $60x - 36$
15) $55x - 25$
16) $-72x + 36$
17) $-42x + 14$
18) $108x - 9$
19) $18x - 12$
20) $16x - 4$
21) $-20x + 40$
22) $-48x + 36$
23) $-45x + 30$
24) $-8x + 10$
25) $22x - 16$
26) $-112x + 70$

Chapter 7:

Equations and Inequalities

Math Topics that you'll learn in this Chapter:

- ✓ One–Step Equations
- ✓ Multi–Step Equations
- ✓ System of Equations
- ✓ Graphing Single–Variable Inequalities
- ✓ One–Step Inequalities
- ✓ Multi–Step Inequalities

One–Step Equations

Solve each equation for x.

1) $x - 15 = 24 \Rightarrow x = $ _____

2) $18 = -6 + x \Rightarrow x = $ _____

3) $19 - x = 8 \Rightarrow x = $ _____

4) $x - 22 = 24 \Rightarrow x = $ _____

5) $24 - x = 17 \Rightarrow x = $ _____

6) $16 - x = 3 \Rightarrow x = $ _____

7) $x + 14 = 12 \Rightarrow x = $ _____

8) $26 + x = 8 \Rightarrow x = $ _____

9) $x + 9 = -18 \Rightarrow x = $ _____

10) $x + 21 = 11 \Rightarrow x = $ _____

11) $17 = -5 + x \Rightarrow x = $ _____

12) $x + 20 = 29 \Rightarrow x = $ _____

13) $x - 13 = 19 \Rightarrow x = $ _____

14) $x + 9 = -17 \Rightarrow x = $ _____

15) $x + 4 = -23 \Rightarrow x = $ _____

16) $16 = -9 + x \Rightarrow x = $ _____

17) $4x = 28 \Rightarrow x = $ _____

18) $21 = -7x \Rightarrow x = $ _____

19) $12x = -12 \Rightarrow x = $ _____

20) $13x = 39 \Rightarrow x = $ _____

21) $8x = -16 \Rightarrow x = $ _____

22) $\frac{x}{2} = -5 \Rightarrow x = $ _____

23) $\frac{x}{9} = 6 \Rightarrow x = $ _____

24) $27 = \frac{x}{5} \Rightarrow x = $ _____

25) $\frac{x}{4} = -3 \Rightarrow x = $ _____

26) $x \div 8 = 7 \Rightarrow x = $ _____

27) $x \div 2 = -3 \Rightarrow x = $ _____

28) $4x = 48 \Rightarrow x = $ _____

29) $9x = 72 \Rightarrow x = $ _____

30) $8x = -32 \Rightarrow x = $ _____

31) $80 = -10x \Rightarrow x = $ _____

Multi –Step Equations

✎ **Solve each equation.**

1) $3x - 8 = 13 \Rightarrow x =$ ___

2) $23 = -(x - 5) \Rightarrow x =$ ___

3) $-(8 - x) = 15 \Rightarrow x =$ ___

4) $29 = -x + 12 \Rightarrow x =$ ___

5) $2(3 - 2x) = 10 \Rightarrow x =$ ___

6) $3x - 3 = 15 \Rightarrow x =$ ___

7) $32 = -x + 15 \Rightarrow x =$ ___

8) $-(10 - x) = -13 \Rightarrow x =$ ___

9) $-4(7 + x) = 4 \Rightarrow x =$ ___

10) $23 = 2x - 8 \Rightarrow x =$ ___

11) $-6(3 + x) = 6 \Rightarrow x =$ ___

12) $-3 = 3x - 15 \Rightarrow x =$ ___

13) $-7(12 + x) = 7 \Rightarrow x =$ ___

14) $8(6 - 4x) = 16 \Rightarrow x =$ ___

15) $18 - 4x = -9 - x \Rightarrow x =$ ___

16) $6(4 - x) = 30 \Rightarrow x =$ ___

17) $15 - 3x = -5 - x \Rightarrow x =$ ___

18) $9(-7 - 3x) = 18 \Rightarrow x =$ ___

19) $16 - 2x = -4 - 7x \Rightarrow x =$ ___

20) $14 - 2x = 14 + x \Rightarrow x =$ ___

21) $21 - 3x = -7 - 10x \Rightarrow x =$ ___

22) $8 - 2x = 11 + x \Rightarrow x =$ ___

23) $10 + 12x = -8 + 6x \Rightarrow x =$ ___

24) $25 + 20x = -5 + 5x \Rightarrow x =$ ___

25) $16 - x = -8 - 7x \Rightarrow x =$ ___

26) $17 - 3x = 13 + x \Rightarrow x =$ ___

27) $22 + 5x = -8 - x \Rightarrow x =$ ___

28) $-9(7 + x) = 9 \Rightarrow x =$ ___

29) $11 + 3x = -4 - 2x \Rightarrow x =$ ___

30) $13 - 2x = 3 - 3x \Rightarrow x =$ ___

31) $19 - x = -1 - 11x \Rightarrow x =$ ___

32) $12 - 2x = -2 - 4x \Rightarrow x =$ ___

System of Equations

✎ **Solve each system of equations.**

1) $-x + y = 2$
   $-2x + y = 3$
   $x =$
   $y =$

2) $-5x + y = -3$
   $3x - 8y = 24$
   $x =$
   $y =$

3) $y = -5$
   $4x - 5y = 13$
   $x =$

4) $3y = -6x + 8$
   $5x - 4y = -3$
   $x =$
   $y =$

5) $10x - 8y = -15$
   $-6x + 4y = 13$
   $x =$
   $y =$

6) $-3x - 4y = 5$
   $x - 2y = 5$
   $x =$
   $y =$

7) $5x - 12y = -19$
   $-6x + 7y = 8$
   $x =$
   $y =$

8) $5x - 7y = -2$
   $-x - 2y = -3$
   $x =$
   $y =$

9) $-x + 3y = 3$
   $-7x + 8y = -5$
   $x =$
   $y =$

10) $-4x + 3y = -18$
    $4x - y = 14$
    $x =$
    $y =$

11) $6x - 7y = -8$
    $-x - 4y = -9$
    $x =$
    $y =$

12) $-3x + 2y = -16$
    $4x - y = 13$
    $x =$
    $y =$

13) $-5x + y = -3$
    $3x - 8y = 24$
    $x =$
    $y =$

14) $3x - 2y = 2$
    $x - y = 2$
    $x =$
    $y =$

15) $4x + 7y = 2$
    $6x + 7y = 10$
    $x =$
    $y =$

16) $5x + 7y = 18$
    $-3x + 7y = -22$
    $x =$
    $y =$

Graphing Single–Variable Inequalities

✍ *Graph each inequality.*

1) $x < 6$

2) $x \geq 1$

3) $x \geq -6$

4) $x \leq -2$

5) $x > -1$

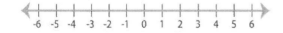

6) $3 > x$

7) $2 \leq x$

8) $x > 0$

9) $-3 \leq x$

10) $-4 \leq x$

11) $x \leq 5$

12) $0 \leq x$

13) $-5 \leq x$

14) $x > -6$

One–Step Inequalities

✎ *Solve each inequality for x.*

1) $x - 10 < 22 \Rightarrow$ _____

2) $18 \leq -4 + x \Rightarrow$ _____

3) $x - 33 > 8 \Rightarrow$ _____

4) $x + 22 \geq 24 \Rightarrow$ _____

5) $x - 24 > 17 \Rightarrow$ _____

6) $x + 5 \geq 3 \Rightarrow x$_____

7) $x + 14 < 12 \Rightarrow$ _____

8) $26 + x \leq 8 \Rightarrow$ _____

9) $x + 9 \geq -18 \Rightarrow$ _____

10) $x + 24 < 11 \Rightarrow$ _____

11) $17 \leq -5 + x \Rightarrow$ _____

12) $x + 25 > 29 \Rightarrow x$_____

13) $x - 17 \geq 19 \Rightarrow$ _____

14) $x + 8 > -17 \Rightarrow$ _____

15) $x + 8 < -23 \Rightarrow$ _____

16) $16 \leq -5 + x \Rightarrow$ _____

17) $4x \leq 12 \Rightarrow$ _____

18) $28 \geq -7x \Rightarrow$ _____

19) $2x > -14 \Rightarrow$ _____

20) $13x \leq 39 \Rightarrow$ _____

21) $-8x > -16 \Rightarrow$ _____

22) $\frac{x}{2} < -6 \Rightarrow$ _____

23) $\frac{x}{6} > 6 \Rightarrow$ _____

24) $27 \leq \frac{x}{4} \Rightarrow$ _____

25) $\frac{x}{8} < -3 \Rightarrow$ _____

26) $6x \geq 18 \Rightarrow$ _____

27) $5x \geq -25 \Rightarrow$ _____

28) $4x > 48 \Rightarrow$ _____

29) $8x \leq 72 \Rightarrow$ _____

30) $-4x < -32 \Rightarrow$ _____

31) $40 > -10x \Rightarrow$ _____

Multi–Step Inequalities

✎ **Solve each inequality.**

1) $2x - 8 \leq 8 \rightarrow$ _____

2) $3 + 2x \geq 17 \rightarrow$ _____

3) $5 + 3x \geq 26 \rightarrow$ _____

4) $2x - 8 \leq 14 \rightarrow$ _____

5) $3x - 4 \leq 23 \rightarrow$ _____

6) $7x - 5 \leq 51 \rightarrow$ _____

7) $4x - 9 \leq 27 \rightarrow$ _____

8) $6x - 11 \leq 13 \rightarrow$ _____

9) $5x - 7 \leq 33 \rightarrow$ _____

10) $6 + 2x \geq 28 \rightarrow$ _____

11) $8 + 3x \geq 35 \rightarrow$ _____

12) $4 + 6x < 34 \rightarrow$ _____

13) $3 + 2x \geq 53 \rightarrow$ _____

14) $7 - 6x > 56 + x \rightarrow$ _____

15) $9 + 4x \geq 39 + 2x \rightarrow$ _____

16) $3 + 5x \geq 43 \rightarrow$ _____

17) $4 - 7x < 60 \rightarrow$ _____

18) $11 - 4x \geq 55 \rightarrow$ _____

19) $12 + x \geq 48 - 2x \rightarrow$ _____

20) $10 - 10x \leq -20 \rightarrow$ _____

21) $5 - 9x \geq -40 \rightarrow$ _____

22) $8 - 7x \geq 36 \rightarrow$ _____

23) $5 + 11x < 69 + 3x \rightarrow$ _____

24) $6 + 8x < 28 - 3x \rightarrow$ _____

25) $9 + 11x < 57 - x \rightarrow$ _____

26) $3 + 10x \geq 45 - 4x \rightarrow$ _____

www.EffortlessMath.com

Answers – Chapter 7

One–Step Equations

1) $x = 39$
2) $x = 24$
3) $x = 11$
4) $x = 46$
5) $x = 7$
6) $x = 13$
7) $x = 26$
8) $x = -18$
9) $x = -27$
10) $x = -10$
11) $x = 22$
12) $x = 9$
13) $x = 32$
14) $x = -26$
15) $x = -19$
16) $x = 25$
17) $x = 7$
18) $x = -3$
19) $x = -1$
20) $x = 3$
21) $x = -2$
22) $x = -10$
23) $x = 54$
24) $x = 135$
25) $x = -12$
26) $x = 56$
27) $x = -6$
28) $x = 12$
29) $x = 8$
30) $x = -4$
31) $x = -8$

Multi–Step Equations

1) $x = 7$
2) $x = -18$
3) $x = 23$
4) $x = -17$
5) $x = -1$
6) $x = 6$
7) $x = -17$
8) $x = -3$
9) $x = -8$
10) $x = 15$
11) $x = -4$
12) $x = 4$
13) $x = -13$
14) $x = 1$
15) $x = 9$
16) $x = -1$
17) $x = 10$
18) $x = -3$
19) $x = -4$
20) $x = 0$
21) $x = -4$
22) $x = -1$
23) $x = -3$
24) $x = -2$
25) $x = -4$
26) $x = 1$
27) $x = -5$
28) $x = -8$
29) $x = -3$
30) $x = -10$
31) $x = -2$
32) $x = -7$

System of Equations

1) $x = -1, y = 1$
2) $x = 0, y = -3$
3) $x = -3$
4) $x = 1, y = 2$
5) $x = -\frac{11}{2}, y = -5$
6) $x = 1, y = -2$
7) $x = 1, y = 2$
8) $x = 1, y = 1$
9) $x = 3, y = 2$
10) $x = 3, y = -2$
11) $x = 1, y = 2$
12) $x = 2, y = -5$
13) $x = 0, y = -3$
14) $x = -2, y = -4$
15) $x = 4, y = -2$
16) $x = 5, y = -1$

Graphing Single–Variable Inequalities

1) $x < 6$

2) $x \geq 1$

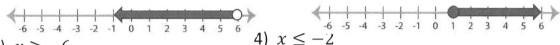

3) $x \geq -6$

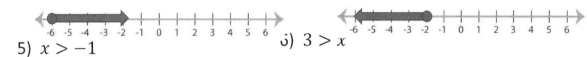

4) $x \leq -2$

5) $x > -1$

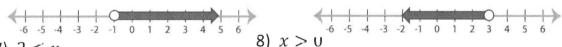

6) $3 > x$

7) $2 \leq x$

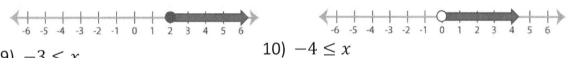

8) $x > 0$

9) $-3 \leq x$

10) $-4 \leq x$

11) $x \leq 5$

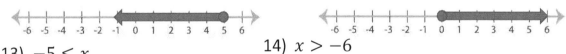

12) $0 \leq x$

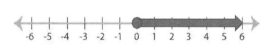

13) $-5 \leq x$

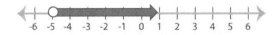

14) $x > -6$

www.EffortlessMath.com

One–Step Inequalities

1) $x < 32$
2) $22 \leq x$
3) $41 \leq x$
4) $x \geq 2$
5) $x > 41$
6) $x \geq -2$
7) $x < -2$
8) $x \leq -18$
9) $x \geq -27$
10) $x < -13$
11) $22 \leq x$
12) $x > 4$
13) $x \geq 36$
14) $x > -25$
15) $x < -31$
16) $21 \leq x$
17) $x \leq 3$
18) $-4 \leq x$
19) $x > -7$
20) $x \leq 3$
21) $x < 2$
22) $x < -12$
23) $x > 36$
24) $108 \leq x$
25) $x < -24$
26) $x \geq 3$
27) $x \geq -5$
28) $x > 12$
29) $x \leq 9$
30) $x > 8$
31) $-4 < x$

Multi–Step Inequalities

1) $x \leq 8$
2) $x \geq 7$
3) $x \geq 7$
4) $x \leq 11$
5) $x \leq 9$
6) $x \leq 8$
7) $x \leq 9$
8) $x \leq 4$
9) $x \leq 8$
10) $x \geq 11$
11) $x \geq 9$
12) $x < 5$
13) $x \geq 25$
14) $x < -7$
15) $x \geq 15$
16) $x \geq 8$
17) $x > -8$
18) $x \leq -11$
19) $x \geq 12$
20) $x \geq 3$
21) $x \leq 5$
22) $x \leq -4$
23) $x < 8$
24) $x < 2$
25) $x < 4$
26) $x \geq 3$

Chapter 8:

Lines and Slope

Math Topics that you'll learn in this Chapter:

- ✓ Finding Slope
- ✓ Graphing Lines Using Slope–Intercept Form
- ✓ Writing Linear Equations
- ✓ Graphing Linear Inequalities
- ✓ Finding Midpoint
- ✓ Finding Distance of Two Points

Finding Slope

✎ **Find the slope of each line.**

1) $y = x - 5$, Slope =

2) $y = -3x + 2$, Slope =

3) $y = -x - 1$, Slope =

4) $y = -x - 9$, Slope =

5) $y = 5 + 2x$, Slope =

6) $y = 1 - 8x$, Slope =

7) $y = -4x + 3$, Slope =

8) $y = -9x + 8$, Slope =

9) $y = -2x + 4$, Slope =

10) $y = 9x - 8$, Slope =

11) $y = \frac{1}{2}x + 4$, Slope =

12) $y = -\frac{2}{5}x + 7$, Slope =

13) $-x + 3y = 5$, Slope =

14) $4x + 4y = 6$, Slope =

15) $6y - 2x = 10$, Slope =

16) $3y - x = 2$, Slope =

✎ **Find the slope of the line through each pair of points.**

17) $(4, 4), (8, 12)$, Slope =

23) $(8, 4), (9, 6)$, Slope =

18) $(-2, 4), (0, 6)$, Slope =

24) $(10, -1), (7, 8)$, Slope =

19) $(6, -2), (2, 6)$, Slope =

25) $(14, -7), (13, -6)$, Slope =

20) $(-4, -2), (0, 6)$, Slope =

26) $(10, 7), (8, 1)$, Slope =

21) $(6, 2), (3, 5)$, Slope =

27) $(5, 1), (8, 10)$, Slope =

22) $(-5, 1), (-1, 9)$, Slope =

28) $(9, -10), (8, 12)$, Slope =

Graphing Lines Using Slope–Intercept Form

✎ *Sketch the graph of each line.*

1) $y = -x + 1$

2) $y = 2x - 3$

3) $y = -x + 2$

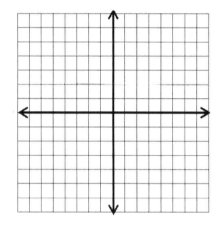
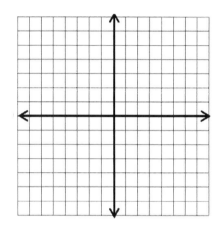
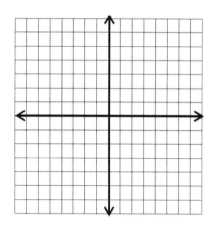

4) $y = x + 1$

5) $y = 2x - 4$

6) $y = -\frac{1}{2}x + 1$

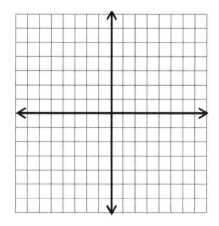
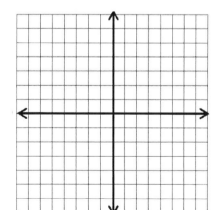
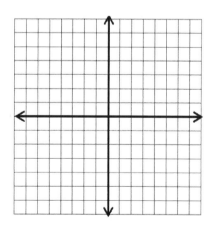

Writing Linear Equations

✎ *Write the equation of the line through the given points.*

1) through: $(1, -2), (2, 4)$

   $y =$

2) through: $(-2, 3), (1, 6)$

   $y =$

3) through: $(-1, 2), (3, 6)$

   $y =$

4) through: $(8, 5), (5, 2)$

   $y =$

5) through: $(7, -10), (2, 10)$

   $y =$

6) through: $(7, 2), (6, 1)$

   $y =$

7) through: $(6, -1), (4, 1)$

   $y =$

8) through: $(-2, 8), (-4, -6)$

   $y =$

9) through: $(-2, 5), (-3, 4)$

   $y =$

10) through: $(6, 8), (8, -6)$

    $y =$

11) through: $(-2, 5), (-4, -3)$

    $y =$

12) through: $(8, 8), (4, -8)$

    $y =$

13) through: $(7, -4)$, Slope: -1

    $y =$

14) through: $(4, -10)$, Slope: -2

    $y =$

15) through: $(6, 10)$, Slope: 9

    $y =$

16) through: $(-6, 8)$, Slope: -2

    $y =$

✎ *Solve each problem.*

17) What is the equation of a line with slope 8 and intercept 5? _____

18) What is the equation of a line with slope 4 and intercept 10? _____

19) What is the equation of a line with slope 9 and passes through point $(5, 23)$?

_____

20) What is the equation of a line with slope -7 and passes through point $(-3, 18)$?

_____

Finding Midpoint

✏️ *Find the midpoint of the line segment with the given endpoints.*

1) $(2, 2), (0, 4)$,

   $midpoint = (\_\_, \_\_)$

2) $(3, 3), (-1, 5)$,

   $midpoint = (\_\_, \_\_)$

3) $(2, -1), (0, 5)$,

   $midpoint = (\_\_, \_\_)$

4) $(-3, 7), (-1, 5)$,

   $midpoint = (\_\_, \_\_)$

5) $(5, -2), (9, -6)$,

   $midpoint = (\_\_, \_\_)$

6) $(-6, -3), (4, -7)$,

   $midpoint = (\_\_, \_\_)$

7) $(7, 0), (-7, 8)$,

   $midpoint = (\_\_, \_\_)$

8) $(-8, 4), (-4, 0)$,

   $midpoint = (\_\_, \_\_)$

9) $(-3, 6), (9, -8)$,

   $midpoint = (\_\_, \_\_)$

10) $(6, 8), (6, -6)$,

    $midpoint = (\_\_, \_\_)$

11) $(6, 7), (-8, 5)$,

    $midpoint = (\_\_, \_\_)$

12) $(9, 3), (-3, -9)$,

    $midpoint = (\_\_, \_\_)$

13) $(-6, 12), (-4, 6)$,

    $midpoint = (\_\_, \_\_)$

14) $(10, 7), (8, -3)$,

    $midpoint = (\_\_, \_\_)$

15) $(13, 7), (-5, 3)$,

    $midpoint = (\_\_, \_\_)$

16) $(-9, -4), (-5, 8)$,

    $midpoint = (\_\_, \_\_)$

17) $(11, 7), (5, 13)$,

    $midpoint = (\_\_, \_\_)$

18) $(-7, -10), (11, -2)$,

    $midpoint = (\_\_, \_\_)$

19) $(10, 15), (-4, 9)$,

    $midpoint = (\_\_, \_\_)$

20) $(11, -4), (7, 12)$,

    $midpoint = (\_\_, \_\_)$

Finding Distance of Two Points

✎ *Find the distance of each pair of points.*

1) $(1, 9), (5, 6)$,

   Distance = ____

2) $(-4, 5), (8, 10)$,

   Distance = ____

3) $(5, -2), (-3, 4)$,

   Distance = ____

4) $(-3, 0), (3, 8)$,

   Distance = ____

5) $(-5, 3), (4, -9)$,

   Distance = ____

6) $(-7, -5), (5, 0)$,

   Distance = ____

7) $(4, 3), (-4, -12)$,

   Distance = ____

8) $(10, 1), (-5, -19)$,

   Distance = ____

9) $(3, 3), (-1, 5)$,

   Distance = ____

10) $(2, -1), (10, 5)$,

    Distance = ____

11) $(-3, 7), (-1, 4)$,

    Distance = ____

12) $(5, -2), (9, -5)$,

    Distance = ____

13) $(-8, 4), (4, 9)$,

    Distance = ____

14) $(6, 8), (6, -6)$,

    Distance = ____

15) $(9, 3), (-3, -2)$,

    Distance = ____

16) $(-4, 12), (-4, 6)$,

    Distance = ____

17) $(-9, -4), (-4, 8)$,

    Distance = ____

18) $(11, 7), (3, 22)$,

    Distance = ____

Answers – Chapter 8

Finding Slope

1) 1
2) −3
3) −1
4) −1
5) 2
6) −8
7) −4
8) −9
9) −2
10) 9
11) $\frac{1}{2}$
12) $-\frac{2}{5}$
13) $\frac{1}{3}$
14) −1
15) $\frac{1}{3}$
16) $\frac{1}{3}$

17) 2
18) 1
19) −2
20) 2
21) −1
22) 2
23) 2
24) −3
25) −1
26) 3
27) 3
28) −22

Graphing Lines Using Slope–Intercept Form

1) $y = -x + 1$

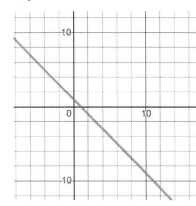

2) $y = 2x - 3$

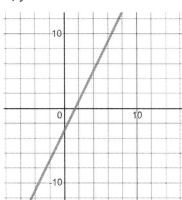

3) $y = -x + 2$

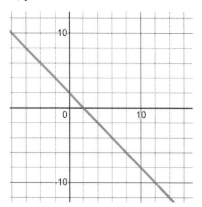

4) $y = x + 1$

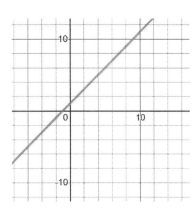

5) $y = 2x - 4$

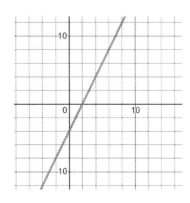

6) $y = -\frac{1}{2}x + 1$

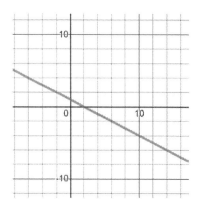

Writing Linear Equations

1) $y = 6x - 8$
2) $y = x + 5$
3) $y = x + 3$
4) $y = x - 3$
5) $y = -4x + 18$
6) $y = x - 5$
7) $y = -x + 5$
8) $y = 7x + 22$
9) $y = x + 7$
10) $y = -7x + 50$
11) $y = 4x + 13$
12) $y = 4x - 24$
13) $y = -x + 3$
14) $y = -2x - 2$
15) $y = 9x - 44$
16) $y = -2x - 4$
17) $y = 8x + 5$
18) $y = 4x + 10$
19) $y = 9x - 22$
20) $y = -7x - 3$

Finding Midpoint

1) $midpoint = (1, 3)$
2) $midpoint = (1, 4)$
3) $midpoint = (1, 2)$
4) $midpoint = (-2, 6)$
5) $midpoint = (7, -4)$
6) $midpoint = (-1, -5)$
7) $midpoint = (0, 4)$
8) $midpoint = (-6, 2)$
9) $midpoint = (3, -1)$
10) $midpoint = (6, 1)$
11) $midpoint = (-1, 6)$
12) $midpoint = (3, -3)$
13) $midpoint = (-5, 9)$
14) $midpoint = (9, 2)$
15) $midpoint = (4, 5)$
16) $midpoint = (-7, 2)$
17) $midpoint = (8, 10)$
18) $midpoint = (2, -6)$
19) $midpoint = (3, 12)$
20) $midpoint = (9, 4)$

Finding Distance of Two Points

1) Distance $= 5$
2) Distance $= 13$
3) Distance $= 10$
4) Distance $= 10$
5) Distance $= 15$
6) Distance $= 13$
7) Distance $= 17$
8) Distance $= 25$
9) Distance $= \sqrt{20} = 2\sqrt{5}$
10) Distance $= 10$
11) Distance $= 5$
12) Distance $= 5$
13) Distance $= 13$
14) Distance $= 14$
15) Distance $= 13$
16) Distance $= 6$
17) Distance $= 13$
18) Distance $= 17$

Chapter 9:

Exponents and Variables

Math Topics that you'll learn in this Chapter:

- ✓ Multiplication Property of Exponents
- ✓ Division Property of Exponents
- ✓ Powers of Products and Quotients
- ✓ Zero and Negative Exponents
- ✓ Negative Exponents and Negative Bases
- ✓ Scientific Notation
- ✓ Radicals

Multiplication Property of Exponents

✎ *Simplify and write the answer in exponential form.*

1) $2 \times 2^2 =$

2) $5^3 \times 5 =$

3) $3^2 \times 3^2 =$

4) $4^2 \times 4^2 =$

5) $7^3 \times 7^2 \times 7 =$

6) $2 \times 2^2 \times 2^2 =$

7) $5^3 \times 5^2 \times 5 \times 5 =$

8) $2x \times x =$

9) $x^3 \times x^2 =$

10) $x^4 \times x^4 =$

11) $x^2 \times x^2 \times x^2 =$

12) $6x \times 6x =$

13) $2x^2 \times 2x^2 =$

14) $3x^2 \times x =$

15) $4x^4 \times 4x^4 \times 4x^4 =$

16) $2x^2 \times x^2 =$

17) $x^4 \times 3x =$

18) $x \times 2x^2 =$

19) $5x^4 \times 5x^4 =$

20) $2yx^2 \times 2x =$

21) $3x^4 \times y^2 x^4 =$

22) $y^2 x^3 \times y^5 x^2 =$

23) $4yx^3 \times 2x^2 y^3 =$

24) $6x^2 \times 6x^3 y^4 =$

25) $3x^4 y^5 \times 7x^2 y^3 =$

26) $7x^2 y^5 \times 9xy^3 =$

27) $7xy^4 \times 4x^3 y^3 =$

28) $3x^5 y^3 \times 8x^2 y^3 =$

29) $3x \times y^5 x^3 \times y^4 =$

30) $yx^2 \times 2y^2 x^2 \times 2xy =$

31) $4yx^4 \times 5y^5 x \times xy^3 =$

32) $7x^2 \times 10x^3 y^3 \times 8yx^4 =$

Division Property of Exponents

✏️ **Simplify and write the answer.**

1) $\dfrac{2^2}{2^3} =$

2) $\dfrac{2^4}{2^2} =$

3) $\dfrac{5^5}{5} =$

4) $\dfrac{3}{3^5} =$

5) $\dfrac{x}{x^3} =$

6) $\dfrac{3 \times 3^3}{3^2 \times 3^4} =$

7) $\dfrac{5^8}{5^3} =$

8) $\dfrac{5 \times 5^6}{5^2 \times 5^7} =$

9) $\dfrac{3^4 \times 3^7}{3^2 \times 3^8} =$

10) $\dfrac{5x}{10x^3} =$

11) $\dfrac{5x^3}{2x^5} =$

12) $\dfrac{18x^3}{14x^6} =$

13) $\dfrac{12x^3}{8xy^8} =$

14) $\dfrac{24xy^3}{4x^4y^2} =$

15) $\dfrac{21\ ^3y^9}{7xy^5} =$

16) $\dfrac{36\ ^2y^9}{4x^3} =$

17) $\dfrac{12x^4y^4}{10x^6y^7} =$

18) $\dfrac{12y^2x^{12}}{20yx^8} =$

19) $\dfrac{16x^4y}{9x^8y^2} =$

20) $\dfrac{5x^8y^2}{20\ ^5y^5} =$

www.EffortlessMath.com

Powers of Products and Quotients

✍ *Simplify and write the answer.*

1) $(4^2)^2 =$

2) $(6^2)^3 =$

3) $(2 \times 2^3)^4 =$

4) $(4 \times 4^4)^2 =$

5) $(3^3 \times 3^2)^3 =$

6) $(5^4 \times 5^5)^2 =$

7) $(2 \times 2^4)^2 =$

8) $(2x^6)^2 =$

9) $(11x^5)^2 =$

10) $(4x^2 y^4)^4 =$

11) $(2x^4 y^4)^3 =$

12) $(3x^2 y^2)^2 =$

13) $(3x^4 y^3)^4 =$

14) $(2x^6 y^8)^2 =$

15) $(12x^3 x)^3 =$

16) $(5x^9 x^6)^3 =$

17) $(5x^{10} y^3)^3 =$

18) $(14x^3 x^3)^2 =$

19) $(3x^3 . 5x)^2 =$

20) $(10x^{11} y^3)^2 =$

21) $(9x^7 y^5)^2 =$

22) $(4x^4 y^6)^5 =$

23) $(3x . 4y^3)^2 =$

24) $\left(\dfrac{6x}{x^2}\right)^2 =$

25) $\left(\dfrac{x^5 y^5}{x^2 y^2}\right)^3 =$

26) $\left(\dfrac{24x}{4x^6}\right)^2 =$

27) $\left(\dfrac{x^5}{x^7 y^2}\right)^2 =$

28) $\left(\dfrac{xy^2}{x^2 y^3}\right)^3 =$

29) $\left(\dfrac{4xy^4}{x^5}\right)^2 =$

30) $\left(\dfrac{xy^4}{5xy^2}\right)^3 =$

Zero and Negative Exponents

✎ *Evaluate the following expressions.*

1) $1^{-1} =$

2) $2^{-2} =$

3) $0^{15} =$

4) $1^{-10} =$

5) $8^{-1} =$

6) $8^{-2} =$

7) $2^{-4} =$

8) $10^{-2} =$

9) $9^{-2} =$

10) $3^{-3} =$

11) $7^{-3} =$

12) $3^{-4} =$

13) $6^{-3} =$

14) $5^{-3} =$

15) $22^{-1} =$

16) $4^{-4} =$

17) $5^{-4} =$

18) $15^{-2} =$

19) $4^{-5} =$

20) $9^{-3} =$

21) $3^{-5} =$

22) $5^{-4} =$

23) $12^{-3} =$

24) $15^{-3} =$

25) $20^{-3} =$

26) $50^{-2} =$

27) $18^{-3} =$

28) $24^{-2} =$

29) $30^{-3} =$

30) $10^{-5} =$

31) $\left(\frac{1}{8}\right)^{-1} =$

32) $\left(\frac{1}{5}\right)^{-2} =$

33) $\left(\frac{1}{7}\right)^{-2} =$

34) $\left(\frac{2}{3}\right)^{-2} =$

35) $\left(\frac{1}{5}\right)^{-3} =$

36) $\left(\frac{3}{4}\right)^{-2} =$

37) $\left(\frac{2}{5}\right)^{-2} =$

38) $\left(\frac{1}{2}\right)^{-8} =$

39) $\left(\frac{2}{5}\right)^{-3} =$

40) $\left(\frac{3}{7}\right)^{-2} =$

41) $\left(\frac{5}{6}\right)^{-3} =$

42) $\left(\frac{4}{9}\right)^{-2} =$

Negative Exponents and Negative Bases

✎ *Simplify and write the answer.*

1) $-3^{-1} =$

2) $-5^{-2} =$

3) $-2^{-4} =$

4) $-x^{-3} =$

5) $2x^{-1} =$

6) $-4x^{-3} =$

7) $-12x^{-5} =$

8) $-5x^{-2}y^{-3} =$

9) $20x^{-4}y^{-1} =$

10) $14a^{-6}b^{-7} =$

11) $-12x^2y^{-3} =$

12) $-\dfrac{25}{x^{-6}} =$

13) $-\dfrac{2x}{a^{-4}} =$

14) $\left(-\dfrac{1}{3x}\right)^{-2} =$

15) $\left(-\dfrac{3}{4x}\right)^{-2} =$

16) $-\dfrac{9}{a^{-7}b^{-2}} =$

17) $-\dfrac{5x}{x^{-3}} =$

18) $-\dfrac{a^{-3}}{b^{-2}} =$

19) $-\dfrac{8}{x^{-3}} =$

20) $\dfrac{5b}{-9c^{-4}} =$

21) $\dfrac{9ab}{a^{-3}b^{-1}} =$

22) $-\dfrac{15a^{-2}}{30b^{-3}} =$

23) $\dfrac{4ab^{-2}}{-3c^{-2}} =$

24) $\left(\dfrac{3a}{2c}\right)^{-2} =$

25) $\left(-\dfrac{5x}{3yz}\right)^{-3} =$

26) $\dfrac{11ab^{-2}}{-3c^{-2}} =$

27) $\left(-\dfrac{x^3}{x^4}\right)^{-2} =$

28) $\left(-\dfrac{x^{-2}}{3x^2}\right)^{-3} =$

Scientific Notation

✎ *Write each number in scientific notation.*

1) $0.113 =$

2) $0.02 =$

3) $7.5 =$

4) $20 =$

5) $60 =$

6) $0.004 =$

7) $78 =$

8) $1,600 =$

9) $1,450 =$

10) $31,000 =$

11) $2,000,000 =$

12) $0.0000003 =$

13) $554,000 =$

14) $0.000725 =$

15) $0.00034 =$

16) $86,000,000 =$

17) $62,000 =$

18) $97,000,000 =$

19) $0.0000045 =$

20) $0.0019 =$

✎ *Write each number in standard notation.*

21) $2 \times 10^{-1} =$

22) $8 \times 10^{-2} =$

23) $1.8 \times 10^3 =$

24) $9 \times 10^{-4} =$

25) $1.7 \times 10^{-2} =$

26) $9 \times 10^3 =$

27) $7 \times 10^5 =$

28) $1.15 \times 10^4 =$

29) $7 \times 10^{-5} =$

30) $8.3 \times 10^{-5} =$

Radicals

✎ **Simplify and write the answer.**

1) $\sqrt{0} =$ ___

2) $\sqrt{1} =$ ___

3) $\sqrt{4} =$ ___

4) $\sqrt{16} =$ ___

5) $\sqrt{9} =$ ___

6) $\sqrt{25} =$ ___

7) $\sqrt{49} =$ ___

8) $\sqrt{36} =$ ___

9) $\sqrt{64} =$ ___

10) $\sqrt{81} =$ ___

11) $\sqrt{121} =$ ___

12) $\sqrt{225} =$ ___

13) $\sqrt{144} =$ ___

14) $\sqrt{100} =$ ___

15) $\sqrt{256} =$ ___

16) $\sqrt{289} =$ ___

17) $\sqrt{324} =$ ___

18) $\sqrt{400} =$ ___

19) $\sqrt{900} =$ ___

20) $\sqrt{529} =$ ___

21) $\sqrt{361} =$ ___

22) $\sqrt{169} =$ ___

23) $\sqrt{196} =$ ___

24) $\sqrt{90} =$ ___

✎ **Evaluate.**

25) $\sqrt{6} \times \sqrt{6} =$

26) $\sqrt{5} \times \sqrt{5} =$

27) $\sqrt{8} \times \sqrt{8} =$

28) $\sqrt{2} + \sqrt{2} =$

29) $\sqrt{8} + \sqrt{8} =$

30) $6\sqrt{5} - 2\sqrt{5} =$

31) $\sqrt{25} \times \sqrt{16} =$

32) $\sqrt{25} \times \sqrt{64} =$

33) $\sqrt{81} \times \sqrt{25} =$

34) $5\sqrt{3} \times 2\sqrt{3} =$

35) $8\sqrt{2} \times 2\sqrt{2} =$

36) $6\sqrt{3} - \sqrt{12} =$

Answers – Chapter 9

Multiplication Property of Exponents

1) 2^3
2) 5^4
3) 3^4
4) 4^4
5) 7^6
6) 2^5
7) 5^7
8) $2x^2$
9) x^5
10) x^8
11) x^6
12) $36x^2$
13) $4x^4$
14) $3x^3$
15) $64x^{12}$
16) $2x^4$
17) $3x^5$
18) $2x^3$
19) $25x^8$
20) $4x^3y$
21) $3x^8y^2$
22) x^5y^7
23) $8x^5y^4$
24) $36x^5y^4$
25) $21x^6y^8$
26) $63x^3y^8$
27) $28x^4y^7$
28) $24x^7y^6$
29) $3x^4y^9$
30) $4x^5y^4$
31) $20x^6y^9$
32) $560x^9y^4$

Division Property of Exponents

1) $\frac{1}{2}$
2) 2^2
3) 5^4
4) $\frac{1}{3^4}$
5) $\frac{1}{x^2}$
6) $\frac{1}{3}$
7) 5^5
8) $\frac{1}{5^2}$
9) 3
10) $\frac{1}{2x^2}$
11) $\frac{5}{2x^2}$
12) $\frac{9}{7x^3}$
13) $\frac{3x^2}{2y^8}$
14) $\frac{6y}{x^3}$
15) $3x^2y^4$
16) $\frac{9y^9}{x}$
17) $\frac{6}{5x^2y^3}$
18) $\frac{3yx^4}{5}$
19) $\frac{16}{9x^4y}$
20) $\frac{x^3}{4y^3}$

Powers of Products and Quotients

1) 4^4
2) 6^6
3) 2^{16}
4) 4^{10}
5) 3^{15}
6) 5^{18}
7) 2^{10}
8) $4x^{12}$
9) $121x^{10}$
10) $256x^8y^{16}$
11) $8x^{12}y^{12}$
12) $9x^4y^4$
13) $81x^{16}y^{12}$
14) $4x^{12}y^{16}$
15) $1,728x^{12}$
16) $125x^{45}$
17) $125x^{30}y^9$
18) $196x^{12}$
19) $225x^8$
20) $100x^{22}y^6$
21) $81x^{14}y^{10}$
22) $1,024x^{20}y^{30}$
23) $144x^2y^6$
24) $\frac{36}{x^2}$
25) x^9y^9
26) $\frac{36}{x^{10}}$

27) $\frac{1}{x^4 y^4}$

28) $\frac{1}{x^3 y^3}$

29) $\frac{16 y^8}{x^8}$

30) $\frac{y^6}{125}$

Zero and Negative Exponents

1) 1
2) $\frac{1}{4}$
3) 0
4) 1
5) $\frac{1}{8}$
6) $\frac{1}{64}$
7) $\frac{1}{16}$
8) $\frac{1}{100}$
9) $\frac{1}{81}$
10) $\frac{1}{27}$
11) $\frac{1}{343}$
12) $\frac{1}{81}$
13) $\frac{1}{216}$
14) $\frac{1}{125}$
15) $\frac{1}{22}$

16) $\frac{1}{256}$
17) $\frac{1}{625}$
18) $\frac{1}{225}$
19) $\frac{1}{1,024}$
20) $\frac{1}{729}$
21) $\frac{1}{243}$
22) $\frac{1}{625}$
23) $\frac{1}{144}$
24) $\frac{1}{3,375}$
25) $\frac{1}{8,000}$
26) $\frac{1}{2,500}$
27) $\frac{1}{5,832}$
28) $\frac{1}{576}$

29) $\frac{1}{27,000}$
30) $\frac{1}{100,000}$
31) 8
32) 25
33) 49
34) $\frac{9}{4}$
35) 125
36) $\frac{64}{27}$
37) $\frac{25}{4}$
38) 256
39) $\frac{125}{8}$
40) $\frac{49}{9}$
41) $\frac{216}{125}$
42) $\frac{81}{16}$

Negative Exponents and Negative Bases

1) $-\frac{1}{3}$
2) $-\frac{1}{25}$
3) $-\frac{1}{16}$
4) $-\frac{1}{x^3}$
5) $\frac{2}{x}$
6) $-\frac{4}{x^3}$
7) $-\frac{12}{x^5}$
8) $-\frac{5}{x^2 y^3}$
9) $\frac{20}{x^4 y}$

10) $\frac{14}{a^6 b^7}$
11) $-\frac{12 x^2}{y^3}$
12) $-25 x^6$
13) $-2 x a^4$
14) $9 x^2$
15) $\frac{16 x^2}{9}$
16) $-9 a^7 b^2$
17) $-5 x^4$
18) $-\frac{b^2}{a^3}$
19) $-8 x^3$
20) $-\frac{5 b c^4}{9}$

21) $9 a^4 b^2$
22) $-\frac{b^3}{2 a^2}$
23) $-\frac{4 a c^2}{3 b^2}$
24) $\frac{4 c^2}{9 a^2}$
25) $-\frac{27 y^3 z^3}{125 x^3}$
26) $-\frac{11 a c^2}{3 b^2}$
27) x^2
28) $-27 x^{12}$

Scientific Notation

1) 1.13×10^{-1}
2) 2×10^{-2}
3) 2.5×10^{0}
4) 2×10^{1}
5) 6×10^{1}
6) 4×10^{-3}
7) 7.8×10^{1}
8) 1.6×10^{3}
9) 1.45×10^{3}
10) 3.1×10^{4}
20) 1.9×10^{-3}
21) $= 0.2$
22) 0.08
23) $1,800$
24) 0.0009
25) 0.017

11) 2×10^{6}
12) 3×10^{-7}
13) 5.54×10^{5}
14) 7.25×10^{-4}
15) 3.4×10^{-4}
16) 8.6×10^{7}
17) 6.2×10^{4}
18) 9.7×10^{7}
19) 4.5×10^{-6}

26) $9,000$
27) $700,000$
28) $11,500$
29) 0.00007
30) 0.000083

Radicals

1) 0
2) 1
3) 2
4) 4
5) 3
6) 5
7) 7
8) 6

9) 8
10) 9
11) 11
12) 15
13) 12
14) 10
15) 16
16) 17

17) 18
18) 20
19) 30
20) 23
21) 19
22) 13
23) 14
24) $3\sqrt{10}$

25) 6
26) 5
27) 8
28) $2\sqrt{2}$

29) $2\sqrt{8} = 4\sqrt{2}$
30) $4\sqrt{5}$
31) 20
32) 40

33) 45
34) 30
35) 32
36) $4\sqrt{3}$

Chapter 10:

Polynomials

Math Topics that you'll learn in this Chapter:

- ✓ Simplifying Polynomials
- ✓ Adding and Subtracting Polynomials
- ✓ Multiplying Monomials
- ✓ Multiplying and Dividing Monomials
- ✓ Multiplying a Polynomial and a Monomial
- ✓ Multiplying Binomials
- ✓ Factoring Trinomials

Simplifying Polynomials

✎ *Simplify each expression.*

1) $2(2x + 2) =$

_____

2) $4(4x - 2) =$

_____

3) $3(5x + 3) =$

_____

4) $6(7x + 5) =$

_____

5) $-3(8x - 7) =$

_____

6) $2x(3x + 4) =$

_____

7) $3x^2 + 3x^2 - 2x^3 =$

_____

8) $2x - x^2 + 6x^3 + 4 =$

_____

9) $5x + 2x^2 - 9x^3 =$

_____

10) $7x^2 + 5x^4 - 2x^3 =$

_____

11) $-3x^2 + 5x^3 + 6x^4 =$

_____

12) $(x - 3)(x - 4) =$

_____

13) $(x - 5)(x + 4) =$

_____

14) $(x - 6)(x - 3) =$

_____

15) $(2x + 5)(x + 8) =$

_____

16) $(3x - 8)(x + 4) =$

_____

17) $-8x^2 + 2x^3 - 10x^4 + 5x =$

_____

18) $11 - 6x^2 + 5x^2 - 12x^3 + 22 =$

_____

19) $2x^2 - 2x + 3x^3 + 12x - 22x =$

_____

20) $11 - 4x^2 + 3x^2 - 7x^3 + 3 =$

_____

21) $2x^5 - x^3 + 8x^2 - 2x^5 =$

_____

22) $(2x^3 - 1) + (3x^3 - 2x^3) =$

_____

www.EffortlessMath.com

Adding and Subtracting Polynomials

Add or subtract expressions.

1) $(x^2 - 3) + (x^2 + 1) =$
_____

2) $(2x^2 - 4) - (2 - 4x^2) =$
_____

3) $(x^3 + 2x^2) - (x^3 + 5) =$
_____

4) $(3x^3 - x^2) + (4x^2 - 7x) =$
_____

5) $(2x^3 + 3x) - (5x^3 + 2) =$
_____

6) $(5x^3 - 2) + (2x^3 + 10) =$
_____

7) $(7x^3 + 5) - (9 - 4x^3) =$
_____

8) $(5x^2 + 3x^3) - (2x^3 + 6) =$
_____

9) $(8x^2 - x) + (4x - 8x^2) =$
_____

10) $(6x + 9x^2) - (5x + 2) =$
_____

11) $(7x^4 - 2x) - (6x - 2x^4) =$
_____

12) $(2x - 4x^3) - (9x^3 + 6x) =$
_____

13) $(8x^3 - 8x^2) - (6x^2 - 3x) =$
_____

14) $(9x^2 - 6) + (5x^2 - 4x^3) =$
_____

15) $(8x^3 + 3x^4) - (x^4 - 3x^3) =$
_____

16) $(-4x^3 - 2x) + (5x - 2x^3) =$
_____

17) $(6x - 4x^4) - (8x^4 + 3x) =$
_____

18) $(7x - 8x^2) - (9x^4 - 3x^2) =$
_____

19) $(9x^3 - 6) + (9x^3 - 5x^2) =$
_____

20) $(5x^3 + x^4) - (8x^4 - 7x^3) =$
_____

Multiplying Monomials

✏️ *Simplify each expression.*

1) $5x^8 \times x^3 =$

2) $5y^5 \times 6y^3 =$

3) $-4z^7 \times 5z^5 =$

4) $7x^5y \times 3xy^2 =$

5) $-6xy^8 \times 3x^5y^3 =$

6) $7a^4b^2 \times 3a^8b =$

7) $5xy^5 \times 3x^3y^4 =$

8) $5p^5q^4 \times (-6pq^4) =$

9) $8s^6t^2 \times 6s^3t^7 =$

10) $(-8x^5y^2) \times 4x^6y^3 =$

11) $9xy^6z \times 3y^4z^2 =$

12) $12x^5y^4 \times 2x^8y =$

13) $4pq^5 \times (-7p^4q^8) =$

14) $9s^4t^2 \times (-5st^5) =$

15) $10p^3q^5 \times (-4p^4q^6) =$

16) $(-5p^2q^4r) \times 7pq^5r^3 =$

17) $(-9a^4b^7c^4) \times (-4a^7b) =$

18) $7u^5v^9 \times (-5u^{12}v^7) =$

19) $5u^3v^9z^2 \times (-4uv^9z) =$

20) $(-9xy^2z^4) \times 2x^2yz^5 =$

21) $8x^3y^2z^5 \times (-9x^4y^2z) =$

22) $6a^8b^8c^{12} \times 9a^7b^5c^8 =$

Multiplying and Dividing Monomials

Simplify each expression.

1) $(8x^3)(2x^2) =$

2) $(4x^6)(5x^4) =$

3) $(-6x^8)(3x^3) =$

4) $(5x^8y^9)(-6x^6y^9) =$

5) $(8x^5y^6)(3x^2y^5) =$

6) $(8yx^2)(7y^5x^3) =$

7) $(4x^2y)(2x^2y^3) =$

8) $(-2x^9y^4)(-9x^6y^8) =$

9) $(-5x^8y^2)(-6x^4y^5) =$

10) $(8x^8y)(-7x^4y^3) =$

11) $(9x^6y^2)(6x^7y^4) =$

12) $(8x^9y^5)(6x^5y^4) =$

13) $(-5x^8y^9)(7x^7y^8) =$

14) $(6x^2y^5)(5x^3y^2) =$

15) $(9x^5y^{12})(4x^7y^9) =$

16) $(-10x^{14}y^8)(2x^7y^5) =$

17) $\dfrac{8x^4y^3}{xy^2} =$

18) $\dfrac{6x^5y^6}{2x^3y} =$

19) $\dfrac{12x^3y^7}{4xy} =$

20) $\dfrac{-20x^8y^9}{5x^5y^4} =$

Multiplying a Polynomial and a Monomial

Find each product.

1) $x(x - 2) =$

2) $2(2 + x) =$

3) $x(x - 1) =$

4) $x(x + 3) =$

5) $2x(x - 2) =$

6) $5(4x + 3) =$

7) $4x(3x - 4) =$

8) $x(5x + 2y) =$

9) $3x(x - 2y) =$

10) $6x(3x - 4y) =$

11) $2x(3x - 8) =$

12) $6x(4x - 6y) =$

13) $3x(4x - 2y) =$

14) $2x(2x - 6y) =$

15) $5x(x^2 + y^2) =$

16) $3x(2x^2 - y^2) =$

17) $7(2x^2 + 9y^2) =$

18) $2x(-2x^2y + 3y) =$

19) $-2(2x^2 - 4xy + 2) =$

20) $5(x^2 - 6xy - 8) =$

Multiplying Binomials

✎ *Find each product.*

1) $(x - 2)(x + 5) =$ _____

2) $(x + 4)(x + 2) =$ _____

3) $(x - 2)(x - 4) =$ _____

4) $(x - 8)(x - 2) =$ _____

5) $(x - 7)(x - 5) =$ _____

6) $(x + 6)(x + 2) =$ _____

7) $(x - 9)(x + 3) =$ _____

8) $(x - 8)(x - 5) =$ _____

9) $(x + 3)(x + 7) =$ _____

10) $(x - 9)(x + 4) =$ _____

11) $(x + 6)(x + 6) =$ _____

12) $(x + 7)(x + 7) =$ _____

13) $(x - 8)(x + 7) =$ _____

14) $(x + 9)(x + 9) =$ _____

15) $(x - 8)(x - 8) =$ _____

16) $(2x - 9)(x + 5) =$ _____

17) $(2x - 3)(x + 4) =$ _____

18) $(2x + 4)(x + 2) =$ _____

19) $(2x + 2)(x + 3) =$ _____

20) $(2x - 4)(2x + 2) =$ _____

Factoring Trinomials

Factor each trinomial.

1) $x^2 + 3x - 10 =$ _____

2) $x^2 + 6x + 8 =$ _____

3) $x^2 - 6x + 8 =$ _____

4) $x^2 - 10x + 16 =$ _____

5) $x^2 - 13x + 40 =$ _____

6) $x^2 + 8x + 12 =$ _____

7) $x^2 - 6x - 27 =$ _____

8) $x^2 - 14x + 48 =$ _____

9) $x^2 + 15x + 56 =$ _____

10) $x^2 - 5x - 36 =$ _____

11) $x^2 + 12x + 36 =$ _____

12) $x^2 + 16x + 63 =$ _____

13) $x^2 + x - 72 =$ _____

14) $x^2 + 18x + 81 =$ _____

15) $x^2 - 16x + 64 =$ _____

16) $x^2 - 18x + 81 =$ _____

17) $2x^2 + 8x + 6 =$ _____

18) $2x^2 + 6x - 8 =$ _____

19) $2x^2 + 12x + 10 =$ _____

20) $4x^2 + 6x - 28 =$ _____

Answers – Chapter 10

Simplifying Polynomials

1) $4x + 4$
2) $16x - 8$
3) $15x + 9$
4) $42x + 30$
5) $-24x + 21$
6) $6x^2 + 8x$
7) $-2x^3 + 6x^2$
8) $6x^3 - x^2 + 2x + 4$
9) $-9x^3 + 2x^2 + 5x$
10) $5x^4 - 2x^3 + 7x^2$
11) $6x^4 + 5x^3 - 3x^2$
12) $x^2 - 7x + 12$
13) $x^2 - x - 20$
14) $x^2 - 9x + 18$
15) $2x^2 + 21x + 40$
16) $3x^2 + 4x - 32$
17) $-10x^4 + 2x^3 - 8x^2 + 5x$
18) $-12x^3 - x^2 + 33$
19) $3x^3 + 2x^2 - 12x$
20) $-7x^3 - x^2 + 14$
21) $-x^3 + 8x^2$
22) $3x^3 - 1$

Adding and Subtracting Polynomials

1) $2x^2 - 2$
2) $6x^2 - 6$
3) $2x^2 - 5$
4) $3x^3 + 3x^2 - 7x$
5) $-3x^3 + 3x^2 - 2$
6) $7x^3 + 8$
7) $11x^3 - 4$
8) $x^3 + 5x^2 - 6$
9) $3x$
10) $9x^2 + x - 2$
11) $9x^4 - 8x$
12) $-13x^3 - 4x$
13) $8x^3 - 14x^2 + 3x$
14) $-4x^3 + 14x^2 - 6$
15) $2x^4 + 11x^3$
16) $-6x^3 + 3x$
17) $-12x^4 + 3x$
18) $-9x^4 - 5x^2 + 7x$
19) $18x^3 - 5x^2 - 6$
20) $-7x^4 + 12x^3$

Multiplying Monomials

1) $5x^{11}$
2) $30y^8$
3) $-20z^{12}$
4) $21x^6y^3$
5) $-18x^6y^{11}$
6) $21a^{12}b^3$
7) $15x^4y^9$
8) $-30p^6q^8$
9) $42s^9t^9$
10) $-32x^{11}y^5$
11) $27xy^{10}z^3$
12) $24x^{13}y^5$
13) $-28p^5q^{13}$
14) $-45s^5t^7$
15) $-40p^7q^{11}$
16) $-35p^3q^9r^4$
17) $36a^{11}b^8c^4$
18) $-35u^{17}v^{16}$
19) $-20u^4v^{18}z^3$
20) $-18x^3y^3z^9$

21) $-72x^7y^4z^6$ 22) $54a^{15}b^{13}c^{20}$

Multiplying and Dividing Monomials

1) $16x^5$
2) $20x^{10}$
3) $-18x^{11}$
4) $-30x^{14}y^{18}$
5) $24x^7y^{11}$
6) $56y^6x^5$
7) $8x^4y^4$
8) $18x^{15}y^{12}$
9) $30x^{12}y^7$
10) $-56x^{12}y^4$
11) $54x^{13}y^6$
12) $48x^{14}y^9$
13) $-35x^{15}y^{17}$
14) $30x^5y^7$
15) $36x^{12}y^{21}$
16) $-20x^{21}y^{13}$
17) $8x^3y$
18) $3x^2y^5$
19) $3x^2y^6$
20) $-4x^3y^5$

Multiplying a Polynomial and a Monomial

1) $x^2 - 2x$
2) $2x + 4$
3) $x^2 - x$
4) $x^2 - 3x$
5) $2x^2 - 4x$
6) $20x + 15$
7) $12x^2 - 16x$
8) $5x^2 + 2xy$
9) $3x^2 - 6xy$
10) $18x^2 - 24xy$
11) $6x^2 - 16x$
12) $24x^2 - 36xy$
13) $12x^2 - 6xy$
14) $4x^2 - 12xy$
15) $5x^3 - 5xy^2$
16) $6x^3 - 3xy^2$
17) $14x^3 + 63y^2$
18) $-4x^3y + 6xy$
19) $-4x^2 + 8xy - 4$
20) $5x^2 - 30xy - 40$

Multiplying Binomials

1) $x^2 + 3x - 10$
2) $x^2 + 6x + 8$
3) $x^2 - 6x + 8$
4) $x^2 - 10x + 16$
5) $x^2 - 12x + 35$
6) $x^2 + 8x + 12$
7) $x^2 - 6x - 27$
8) $x^2 - 13x + 40$
9) $x^2 + 10x + 21$
10) $x^2 - 5x - 36$
11) $x^2 + 12x + 36$
12) $x^2 + 14x + 49$
13) $x^2 - x - 56$
14) $x^2 + 18x + 81$
15) $x^2 - 16x + 64$
16) $x^2 - 4x - 45$
17) $2x^2 + 5x - 12$
18) $2x^2 + 8x + 8$
19) $2x^2 + 8x + 6$
20) $4x^2 - 4x - 8$

Factoring Trinomials

1) $(x-2)(x+5)$
2) $(x+4)(x+2)$
3) $(x-2)(x-4)$
4) $(x-8)(x-2)$
5) $(x-8)(x-5)$
6) $(x+6)(x+2)$
7) $(x-9)(x+3)$
8) $(x-8)(x-6)$
9) $(x+8)(x+7)$
10) $(x-9)(x+4)$
11) $(x+6)(x+6)$
12) $(x+7)(x+9)$
13) $(x-8)(x+9)$
14) $(x+9)(x+9)$
15) $(x-8)(x-8)$
16) $(x-9)(x-9)$
17) $(2x+2)(x+3)$
18) $(2x-2)(x+4)$
19) $(2x+2)(x+5)$
20) $(2x-4)(2x+7)$

Chapter 11:

Geometry and Solid Figures

Math Topics that you'll learn in this Chapter:

- ✓ The Pythagorean Theorem
- ✓ Triangles
- ✓ Polygons
- ✓ Circles
- ✓ Trapezoids
- ✓ Cubes
- ✓ Rectangle Prisms
- ✓ Cylinder

The Pythagorean Theorem

✏️ *Do the following lengths form a right triangle?*

1) _____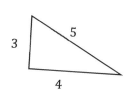

2) _____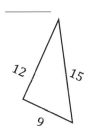

3) _____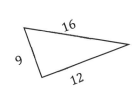

4) _____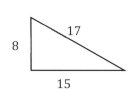

5) _____

6) _____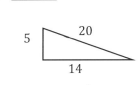

7) _____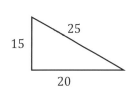

8) _____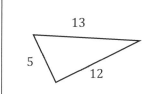

✏️ *Find the missing side.*

9) _____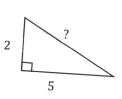

10) _____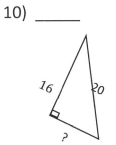

11) _____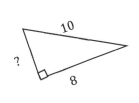

12) _____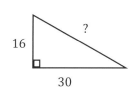

13) _____

14) _____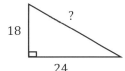

15) _____

16) _____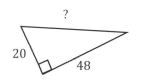

www.EffortlessMath.com

Triangles

✏️ *Find the measure of the unknown angle in each triangle.*

1) _____ 2) _____ 3) _____ 4) _____

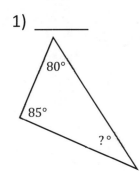 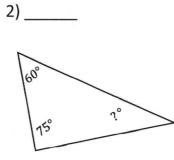 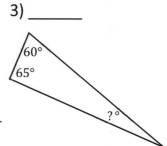 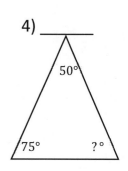

5) _____ 6) _____ 7) _____ 8) _____

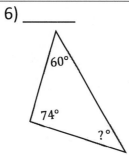 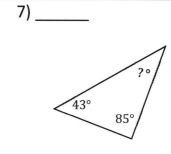 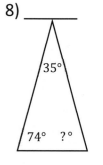

✏️ *Find area of each triangle.*

9) _____ 10) _____ 11) _____ 12) _____

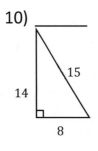 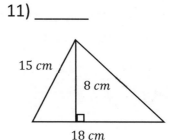 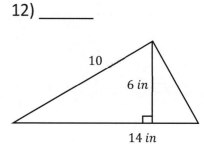

Polygons

✏️ **Find the perimeter of each shape.**

1) (square) _____ 2) _____ 3) _____ 4) (square) _____

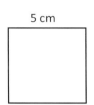
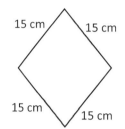
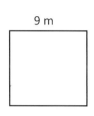

5) (regular hexagon) _____ 6) _____ 7) (parallelogram) _____ 8) (regular hexagon) _____

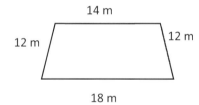
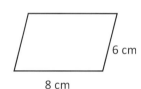

9) _____ 10) _____ 11) _____ 12) (regular hexagon) _____

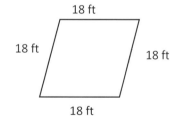
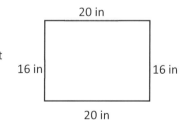
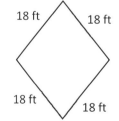
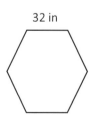

www.EffortlessMath.com   97

Circles

✏️ **Find the Circumference of each circle.** (π = 3.14)

1) _____   2) _____   3) _____   4) _____   5) _____   6) _____

7) _____   8) _____   9) _____   10) _____   11) _____   12) _____

✏️ **Complete the table below.** (π = 3.14)

	Radius	Diameter	Circumference	Area
Circle 1	2 inches	4 inches	12.56 inches	12.56 square inches
Circle 2		8 meters		
Circle 3				113.04 square ft
Circle 4			50.24 miles	
Circle 5		9 km		
Circle 6	7 cm			
Circle 7		10 feet		
Circle 8				615.44 square meters
Circle 9			81.64 inches	
Circle 10	12 feet			

Cubes

✏️ *Find the volume of each cube.*

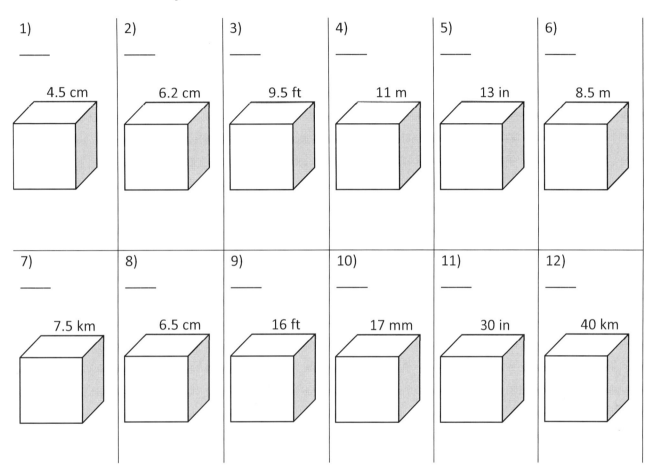

✏️ *Find the surface area of each cube.*

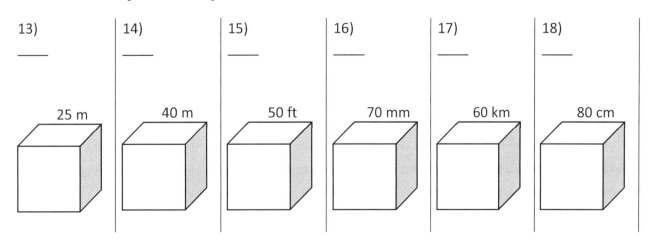

Trapezoids

✎ **Find the area of each trapezoid.**

1) _____

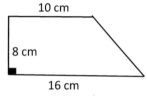

2) _____

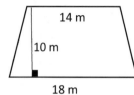

3) _____

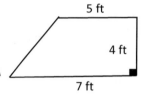

4) _____

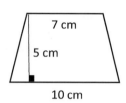

5) _____

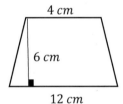

6) _____

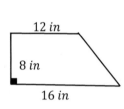

7) _____

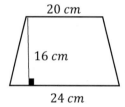

8) _____

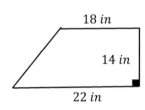

✎ **Solve.**

9) A trapezoid has an area of 80 cm² and its height is 8 cm and one base is 12 cm. What is the other base length? _____

10) If a trapezoid has an area of 120 ft² and the lengths of the bases are 14 ft and 16 ft, find the height. _____

11) If a trapezoid has an area of 160 m² and its height is 10 m and one base is 14 m, find the other base length. _____

12) The area of a trapezoid is 504 ft² and its height is 24 ft. If one base of the trapezoid is 14 ft, what is the other base length? _____

Rectangular Prisms

✎ *Find the volume of each Rectangular Prism.*

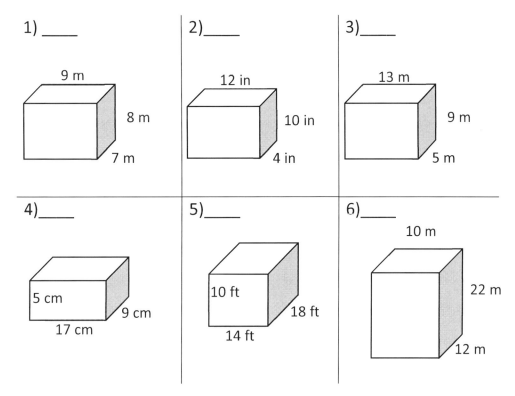

✎ *Find the surface area of each Rectangular Prism.*

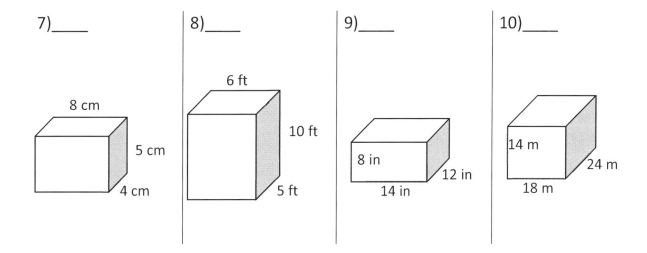

www.EffortlessMath.com

Cylinder

✎ **Find the volume of each Cylinder.** (π = 3.14)

1) _____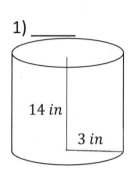

2) _____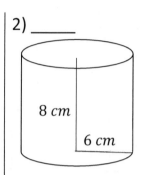

3) _____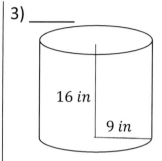

4) _____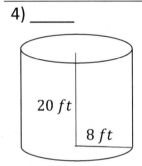

5) _____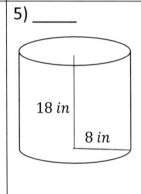

6) _____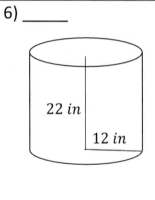

✎ **Find the surface area of each Cylinder.** (π = 3.14)

7) _____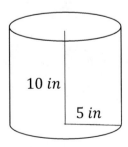

8) _____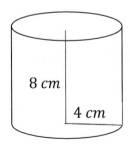

9) _____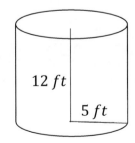

10) _____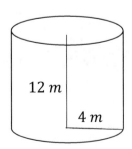

Answers – Chapter 11

The Pythagorean Theorem

1) yes
2) yes
3) no
4) yes
5) no
6) no
7) yes
8) yes
9) 51
10) 12
11) 6
12) 34
13) 26
14) 13
15) 30
16) 52

Triangles

1) 15°
2) 45°
3) 55°
4) 55°
5) 45°
6) 46°
7) 52°
8) 71°
9) 40
10) 56
11) 72 cm^2
12) 42 in^2

Polygons

1) 20 cm
2) 44 m
3) 60 cm
4) 36 m
5) 96 m
6) 56 m
7) 28 cm
8) 120 ft
9) 72 ft
10) 72 in
11) 88 ft
12) 192 in

Circles

1) 43.96 in
2) 75.36 cm
3) 87.92 ft
4) 81.64 m
5) 113.04 cm
6) 94.2 $miles$
7) 119.32 in
8) 138.16 ft
9) 157 m
10) 175.84 m
11) 219.8 in
12) 314 ft

	Radius	Diameter	Circumference	Area
Circle 1	2 inches	4 inches	12.56 inches	12.56 square inches
Circle 2	4 meters	8 meters	25.12 meters	50.24 square meters
Circle 3	6 ft	12 ft	37.68	113.04 square ft
Circle 4	8 miles	16 miles	50.24 miles	200.96 square miles
Circle 5	4.5 km	9 km	28.26 km	63.585 square km
Circle 6	7 cm	14 cm	43.96 cm	153.86 square cm
Circle 7	5 feet	10 feet	31.4 feet	78.5 square feet
Circle 8	14 m	28 m	87.92 m	615.44 square meters
Circle 9	13 in	26 in	81.64 inches	530.66 square inches
Circle 10	12 feet	24 feet	75.36 feet	452.16 square feet

Cubes

1) $91.125\ cm^3$
2) $238.328\ cm^3$
3) $857.375\ ft^3$
4) $1,331\ m^3$
5) $2,197\ in^3$
6) $614.125\ m^3$
7) $421.875\ km^3$
8) $274.625\ cm^3$
9) $4,096\ ft^3$
10) $4,913\ cm^3$
11) $27,000\ in^3$
12) $64,000\ km^3$
13) $3,750\ m^2$
14) $9,600\ m^2$
15) $15,000\ ft^2$
16) $29,400\ mm^2$
17) $21,600\ km^2$
18) $38,400\ cm^2$

Trapezoids

1) $104\ cm^2$
2) $160\ m^2$
3) $224\ ft^2$
4) $324\ cm^2$
5) $288\ cm^2$
6) $414\ in^2$
7) $448\ cm^2$
8) $528\ in^2$
9) $8\ cm$
10) $8\ ft$
11) $18\ m$
12) $28\ ft$

Rectangular Prisms

1) $504\ m^3$
2) $480\ in^3$
3) $585\ m^3$
4) $765\ cm^3$
5) $2,520\ ft^3$
6) $2,640\ m^3$
7) $184\ cm^2$
8) $280\ ft^2$
9) $752\ in^2$
10) $2,040\ m^2$

Cylinder

1) $395.64\ in^3$
2) $904.32\ cm^3$
3) $4,069.44\ in^3$
4) $4,019.2\ ft^3$
5) $3,617.28\ in^3$
6) $9,947.52\ in^3$
7) $471\ in^2$
8) $301.44\ cm^2$
9) $533.8\ ft^2$
10) $401.92\ m^2$

Chapter 12:

Statistics

Math Topics that you'll learn in this Chapter:

- ✓ Mean, Median, Mode, and Range of the Given Data
- ✓ Pie Graph
- ✓ Probability Problems
- ✓ Permutations and Combinations

Mean, Median, Mode, and Range of the Given Data

✎ *Find the values of the Given Data.*

1) 6, 12, 1, 1, 5

   Mode: _____   Range: _____

   Mean: _____   Median: _____

2) 5, 8, 3, 7, 4, 3

   Mode: _____   Range: _____

   Mean: _____   Median: _____

3) 12, 5, 8, 7, 8

   Mode: _____   Range: _____

   Mean: _____   Median: _____

4) 8, 4, 10, 7, 3, 4

   Mode: _____   Range: _____

   Mean: _____   Median: _____

5) 9, 7, 10, 5, 7, 4, 14

   Mode: _____   Range: _____

   Mean: _____   Median: _____

6) 8, 1, 6, 6, 9, 2, 17

   Mode: _____   Range: _____

   Mean: _____   Median: _____

7) 12, 6, 1, 7, 9, 7, 8, 14

   Mode: _____   Range: _____

   Mean: _____   Median: _____

8) 10, 14, 5, 4, 11, 6, 13

   Mode: _____   Range: _____

   Mean: _____   Median: _____

9) 16, 15, 15, 16, 13, 14, 23

   Mode: _____   Range: _____

   Mean: _____   Median: _____

10) 16, 15, 12, 8, 4, 9, 8, 16

   Mode: _____   Range: _____

   Mean: _____   Median: _____

Pie Graph

✎ *The circle graph below shows all Wilson's expenses for last month. Wilson spent $200 on his bills last month.*

Answer following questions based on the Pie graph.

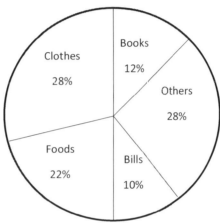

1) How much was Wilson's total expenses last month? _____

2) How much did Wilson spend on his clothes last month? _____

3) How much did Wilson spend for foods last month? _____

4) How much did Wilson spend on his books last month? _____

5) What fraction is Wilson's expenses for his bills and clothes out of his total expenses last month? _____

Probability Problems

1) If there are 10 red balls and 20 blue balls in a basket, what is the probability that Oliver will pick out a red ball from the basket? _____

Gender	Under 45	45 or older	total
Male	12	6	18
Female	5	7	12
Total	17	13	30

2) The table above shows the distribution of age and gender for 30 employees in a company. If one employee is selected at random, what is the probability that the employee selected be either a female under age 45 or a male age 45 or older? _____

3) A number is chosen at random from 1 to 18. Find the probability of not selecting a composite number. (A composite number is a number that is divisible by itself, 1 and at least one other whole number) _____

4) There are 6 blue marbles, 8 red marbles, and 5 yellow marbles in a box. If Ava randomly selects a marble from the box, what is the probability of selecting a red or yellow marble? _____

5) A bag contains 19 balls: three green, five black, eight blue, a brown, a red and one white. If 18 balls are removed from the bag at random, what is the probability that a brown ball has been removed? _____

6) There are only red and blue marbles in a box. The probability of choosing a red marble in the box at random is one fourth. If there are 132 blue marbles, how many marbles are in the box? _____

Permutations and Combinations

✎ *Calculate the value of each.*

1) 5! = ____

2) 6! = ____

3) 8! = ____

4) 5! + 6! = ____

5) 8! + 3! = ____

6) 6! + 7! = ____

7) 8! + 4! = ____

8) 9! − 3! = ____

✎ *Solve each word problems.*

9) Sophia is baking cookies. She uses milk, flour and eggs. How many different orders of ingredients can she try? _____

10) William is planning for his vacation. He wants to go to restaurant, watch a movie, go to the beach, and play basketball. How many different ways of ordering are there for him? _____

11) How many 7-digit numbers can be named using the digits 1, 2, 3, 4, 5, 6 and 7 without repetition? _____

12) In how many ways can 9 boys be arranged in a straight line? _____

13) In how many ways can 10 athletes be arranged in a straight line? _____

14) A professor is going to arrange her 7 students in a straight line. In how many ways can she do this? _____

15) How many code symbols can be formed with the letters for the word BLACK? _____

16) In how many ways a team of 7 basketball players can choose a captain and co-captain? _____

Answers – Chapter 12

Mean, Median, Mode, and Range of the Given Data

1) Mode: 1   Range: 11   Mean: 5   Median: 5
2) Mode: 3   Range: 5   Mean: 5   Median: 4.5
3) Mode: 8   Range: 7   Mean: 8   Median: 8
4) Mode: 4   Range: 7   Mean: 6   Median: 5.5
5) Mode: 7   Range: 10   Mean: 8   Median: 7
6) Mode: 6   Range: 16   Mean: 7   Median: 6
7) Mode: 7   Range: 13   Mean: 8   Median: 7.5
8) Mode: *no mode* Range: 10   Mean: 9   Median: 10
9) Mode: 15 *and* 16 Range: 10 Mean: 16   Median: 15
10) Mode: 8 *and* 16 Range: 12 Mean: 11   Median: 10.5

Pie Graph

1) $2,000
2) $560
3) $440
4) $240
5) $\frac{19}{50}$

Probability Problems

1) $\frac{1}{3}$
2) $\frac{11}{30}$
3) $\frac{7}{18}$
4) $\frac{13}{19}$
5) $\frac{18}{19}$
6) 176

Permutations and Combinations

1) 120
2) 720
3) 40,320
4) 840
5) 40,326
6) 5,760
7) 40,344
8) 362,874
9) *6*
10) *24*
11) 5,040
12) 362,880
13) 3,628,800
14) 5,040
15) 120
16) 42

Chapter 13:

Functions Operations

Math Topics that you'll learn in this Chapter:

- ✓ Function Notation and Evaluation
- ✓ Adding and Subtracting Functions
- ✓ Multiplying and Dividing Functions
- ✓ Composition of Functions

Function Notation and Evaluation

Evaluate each function.

1) $f(x) = x - 1$, find $f(-1)$

   ___

2) $g(x) = x + 3$, find $f(4)$

   ___

3) $h(x) = x + 9$, find $f(3)$

   ___

4) $f(x) = -x - 6$, find $f(5)$

   ___

5) $f(x) = 2x - 7$, find $f(-1)$

   ___

6) $w(x) = -2 - 4x$, find $w(5)$

   ___

7) $g(n) = 6n - 3$, find $g(-2)$

   ___

8) $h(x) = -8x + 12$, find $h(3)$

   ___

9) $k(n) = 14 - 3n$, find $k(3)$

   ___

10) $g(x) = 4x - 4$, find $g(-2)$

    ___

11) $k(n) = 8n - 7$, find $k(4)$

    ___

12) $w(n) = -2n + 14$, find $w(5)$

    ___

13) $h(x) = 5x - 18$, find $h(8)$

    ___

14) $g(n) = 2n^2 + 2$, find $g(5)$

    ___

15) $f(x) = 3x^2 - 13$, find $f(2)$

    ___

16) $g(n) = 5n^2 + 7$, find $g(-3)$

    ___

17) $h(n) = 5n^2 - 10$, find $h(4)$

    ___

18) $g(x) = -3x^2 - 6x$, find $g(2)$

    ___

19) $k(n) = 3n^3 + 2n$, find $k(-5)$

    ___

20) $f(x) = -4x + 12$, find $f(2x)$

    ___

21) $k(a) = 6a + 5$, find $k(a - 1)$

    ___

22) $h(x) = 9x + 3$, find $h(5x)$

    ___

Adding and Subtracting Functions

✎ *Perform the indicated operation.*

1) $f(x) = x + 6$
   $g(x) = 3x + 3$
   Find $(f - g)(2)$
   _____

2) $g(x) = x - 3$
   $f(x) = -x - 4$
   Find $(g - f)(-2)$
   _____

3) $h(t) = 5t + 4$
   $g(t) = 2t + 2$
   Find $(h + g)(-1)$
   _____

4) $g(a) = 3a - 5$
   $f(a) = a^2 + 6$
   Find $(g + f)(3)$
   _____

5) $g(x) = 4x - 5$
   $h(x) = 6x^2 + 5$
   Find $(g - f)(-2)$
   _____

6) $h(x) = x^2 + 3$
   $g(x) = -4x + 1$
   Find $(h + g)(4)$
   _____

7) $f(x) = -2x - 8$
   $g(x) = x^2 + 2$
   Find $(f - g)(6)$
   _____

8) $h(n) = -4n^2 + 9$
   $g(n) = 5n + 6$
   Find $(h - g)(5)$
   _____

9) $g(x) = 3x^2 - 2x - 1$
   $f(x) = 5x + 12$
   Find $(g - f)(a)$
   _____

10) $g(t) = -5t - 8$
    $f(t) = -t^2 + 2t + 12$
    Find $(g + f)(x)$
    _____

www.EffortlessMath.com

Multiplying and Dividing Functions

Perform the indicated operation.

1) $g(x) = x + 2$
   $f(x) = x + 3$
   Find $(g \cdot f)(4)$

   _____

2) $f(x) = 2x$
   $h(x) = -x + 6$
   Find $(f \cdot h)(-2)$

   _____

3) $g(a) = a + 2$
   $h(a) = 2a - 3$
   Find $(g \cdot h)(5)$

   _____

4) $f(x) = 2x + 4$
   $h(x) = 4x - 2$
   Find $\left(\dfrac{f}{h}\right)(2)$

   _____

5) $f(x) = a^2 - 2$
   $g(x) = -4 + 3a$
   Find $\left(\dfrac{f}{g}\right)(2)$

   _____

6) $g(a) = 4a + 6$
   $f(a) = 2a - 8$
   Find $\left(\dfrac{g}{f}\right)(3)$

   _____

7) $g(t) = t^2 + 4$
   $h(t) = 2t - 4$
   Find $(g \cdot h)(-3)$

   _____

8) $g(x) = x^2 + 2x + 5$
   $h(x) = 2x + 3$
   Find $(g \cdot h)(2)$

   _____

9) $g(a) = 2a^2 - 4a + 2$
   $f(a) = 2a^3 - 2$
   Find $\left(\dfrac{g}{f}\right)(4)$

   _____

10) $g(x) = -4x^2 + 5 - 2x$
    $f(x) = x^2 - 2$
    Find $(g \cdot f)(3)$

    _____

Composition of Functions

✎ **Using** $f(x) = x + 4$ **and** $g(x) = 2x$, **find:**

1) $f(g(1)) = $ _____

2) $f(g(-1)) = $ _____

3) $g(f(-2)) = $ _____

4) $g(f(2)) = $ _____

5) $f(g(2)) = $ _____

6) $g(f(3)) = $ _____

✎ **Using** $f(x) = 2x + 5$ **and** $g(x) = x - 2$, **find:**

7) $g(f(2)) = $ _____

8) $g(f(-2)) = $ _____

9) $f(g(5)) = $ _____

10) $f(f(4)) = $ _____

11) $g(f(3)) = $ _____

12) $g(f(-3)) = $ _____

✎ **Using** $f(x) = 4x - 2$ **and** $g(x) = x - 5$, **find:**

13) $g(f(-2)) = $ _____

14) $f(f(4)) = $ _____

15) $f(g(5)) = $ _____

16) $f(f(3)) = $ _____

17) $g(f(-3)) = $ _____

18) $g(g(6)) = $ _____

✎ **Using** $f(x) = 5x + 3$ **and** $g(x) = 2x - 5$, **find:**

19) $f(g(-4)) = $ _____

20) $g(f(6)) = $ _____

21) $f(g(5)) = $ _____

22) $f(f(3)) = $ _____

Answers – Chapter 13

Function Notation and Evaluation

1) -2
2) 7
3) 12
4) -11
5) -9
6) -22
7) -15
8) -12
9) 5
10) -12
11) 25
12) 4
13) 22
14) 52
15) -1
16) 52
17) 70
18) -24
19) -385
20) $-8x + 12$
21) $6a - 1$
22) $45x + 3$

Adding and Subtracting Functions

1) -1
2) -3
3) -1
4) 19
5) -42
6) 4
7) -58
8) -122
9) $3a^2 - 7a - 13$
10) $-x^2 - 3x + 4$

Multiplying and Dividing Functions

1) 42
2) -32
3) 49
4) $\frac{4}{3}$
5) 1
6) -9
7) -130
8) 91
9) $\frac{1}{7}$
10) -259

Composition of Functions

1) $f(g(1)) = 6$
2) $f(g(-1)) = 2$
3) $g(f(-2)) = 4$
4) $g(f(2)) = 12$
5) $f(g(2)) = 8$
6) $g(f(3)) = 14$
7) $g(f(2)) = 7$
8) $g(f(-2)) = -1$
9) $f(g(5)) = 11$
10) $f(f(4)) = 31$
11) $g(f(3)) = 9$
12) $g(f(-3)) = -3$
13) $g(f(-2)) = -15$
14) $f(f(4)) = 54$
15) $f(g(5)) = -2$
16) $f(f(3)) = 38$
17) $g(f(-3)) = -19$
18) $g(g(6)) = -4$
19) $f(g(-4)) = -62$
20) $g(f(6)) = 61$
21) $f(g(5)) = 28$
22) $f(f(3)) = 93$

Chapter 14:

Quadratic

Topics that you'll practice in this chapter:

✓ Solving a Quadratic Equation

✓ Graphing Quadratic Functions

✓ Solving Quadratic Inequalities

✓ Graphing Quadratic Inequalities

Solving a Quadratic Equation

Solve each equation by factoring or using the quadratic formula.

1) $x^2 - 4x - 32 = 0$

_____

2) $x^2 - 2x - 63 = 0$

_____

3) $x^2 + 17x + 72 = 0$

_____

4) $x^2 + 14x + 48 = 0$

_____

5) $x^2 + 5x - 24 = 0$

_____

6) $x^2 + 15x + 36 = 0$

_____

7) $x^2 + 12x - 28 = 0$

_____

8) $x^2 + 6x - 55 = 0$

_____

9) $x^2 + 16x - 105 = 0$

_____

10) $x^2 - 21x + 54 = 0$

_____

11) $x^2 + 8x - 128 = 0$

_____

12) $x^2 + 19x - 150 = 0$

_____

13) $x^2 + 15x - 154 = 0$

_____

14) $2x^2 - 2x - 60 = 0$

_____

15) $2x^2 - 10x - 72 = 0$

_____

16) $4x^2 + 48x + 128 = 0$

_____

17) $4x^2 + 40x + 96 = 0$

_____

18) $2x^2 + 28x + 90 = 0$

_____

19) $9x^2 + 63x + 108 = 0$

_____

20) $4x^2 + 56x + 160 = 0$

_____

Graphing Quadratic Functions

✏️ *Sketch the graph of each function.*

1) $y = (x+1)^2 - 2$

2) $y = (x-1)^2 + 3$

3) $y = x^2 - 4x + 6$

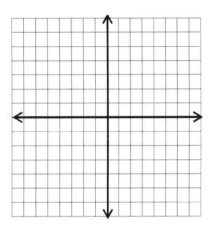
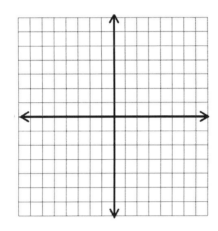
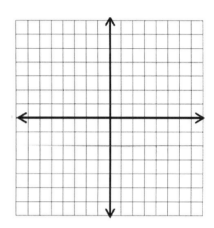

4) $y = x^2 - 6x + 14$

5) $y = x^2 + 12x + 34$

6) $y = 2(x+1)^2 - 4$

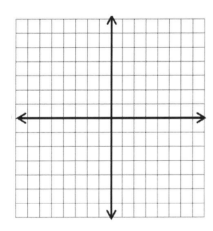
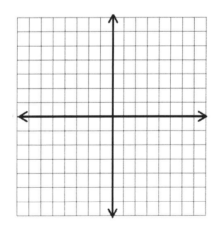
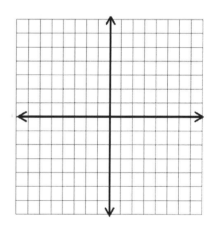

Solving Quadratic Inequalities

✎ *Solve each quadratic inequality.*

1) $x^2 - 4 < 0$

_____

2) $x^2 - 9 > 0$

_____

3) $x^2 - 5x - 6 < 0$

_____

4) $x^2 + 8x - 20 > 0$

_____

5) $x^2 + 10x - 24 \geq 0$

_____

6) $x^2 - 15x + 54 < 0$

_____

7) $x^2 + 17x + 72 \leq 0$

_____

8) $x^2 + 15x + 44 \geq 0$

_____

9) $x^2 + 5x - 50 \geq 0$

_____

10) $x^2 - 18x + 72 < 0$

_____

11) $x^2 - 18x + 45 > 0$

_____

12) $x^2 + 16x - 80 > 0$

_____

13) $x^2 + 9x - 112 \leq 0$

_____

14) $x^2 + 4x - 117 \leq 0$

_____

15) $x^2 + 19x + 88 \geq 0$

_____

16) $x^2 + 26x + 168 \leq 0$

_____

17) $4x^2 + 24x + 32 \leq 0$

_____

18) $4x^2 - 4x - 48 \geq 0$

_____

19) $4x^2 - 16x - 48 \leq 0$

_____

20) $9x^2 - 63x + 108 > 0$

_____

Graphing Quadratic Inequalities

✎ *Sketch the graph of each quadratic inequality.*

1) $y < -2x^2$

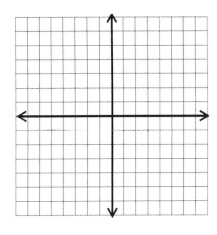

2) $y > 3x^2$

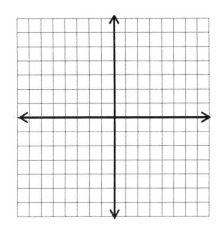

3) $y \geq -3x^2$

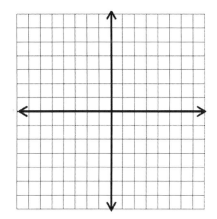

4) $y < x^2 + 1$

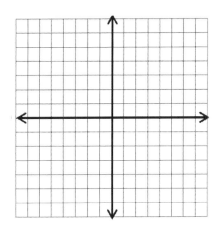

5) $y \geq -x^2 + 2$

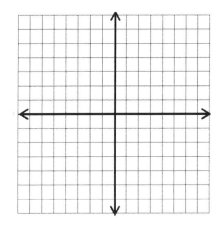

6) $y \leq x^2 - 2x - 3$

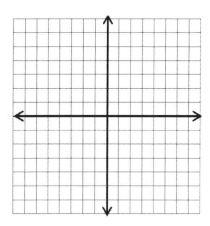

Answers – Chapter 14

Solving a Quadratic Equation

1) $x = 8, x = -4$
2) $x = 9, x = -7$
3) $x = -9, x = -8$
4) $x = -6, x = -8$
5) $x = 3, x = -8$
6) $x = -12, x = -3$
7) $x = 2, x = -14$
8) $x = 5, x = -11$
9) $x = -21, x = 5$
10) $x = 18, x = 3$
11) $x = -16, x = 8$
12) $x = -25, x = 6$
13) $x = -22, x = 7$
14) $x = 6, x = -5$
15) $x = 9, x = -4$
16) $x = -4, x = -8$
17) $x = -4, x = -6$
18) $x = -5, x = -9$
19) $x = -3, x = -4$
20) $x = -4, x = -10$

Graphing Quadratic Functions

1) $y = (x + 1)^2 - 2$

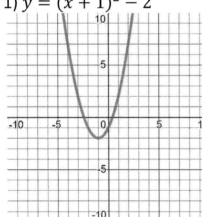

2) $y = (x - 1)^2 + 3$

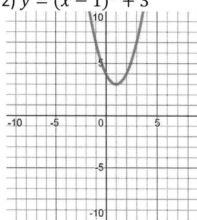

3) $y = x^2 - 4x + 6$

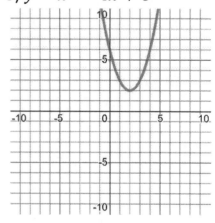

4) $y = x^2 - 6x + 14$

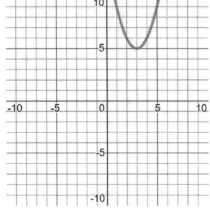

5) $y = x^2 + 12x + 34$

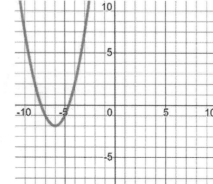

6) $y = 2(x + 1)^2 - 4$

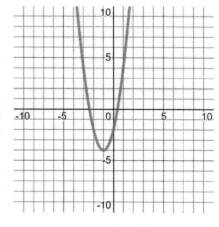

Solving Quadratic Inequalities

1) $-2 < x < 2$
2) $-3 < x < 3$
3) $-1 < x < 6$
4) $x < -10 \text{ or } x > 2$
5) $x \leq -12 \text{ or } x \geq 2$
6) $6 < x < 9$
7) $-9 \leq x \leq -8$
8) $x \leq -11 \text{ or } x \geq -4$
9) $x \leq -10 \text{ or } x \geq 5$
10) $6 \leq x \leq 12$
11) $x < 3 \text{ or } x > 15$
12) $x < -20 \text{ or } x > 4$
13) $-16 \leq x \leq 7$
14) $-13 \leq x \leq 9$
15) $x \leq -11 \text{ or } x \geq -8$
16) $-14 \leq x \leq -12$
17) $-4 \leq x \leq -2$
18) $x \leq -3 \text{ or } x \geq 4$
19) $-2 \leq x \leq 6$
20) $x \leq 3 \text{ or } x \geq 4$

Graphing Quadratic Inequalities

1) $y < -2x^2$

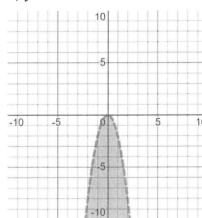

2) $y > 3x^2$

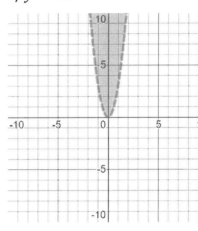

3) $y \geq -3x^2$

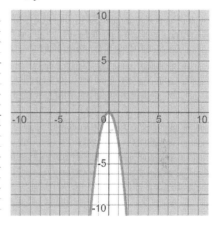

4) $y < x^2 + 1$

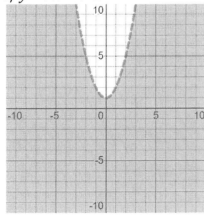

5) $y \geq -x^2 + 2$

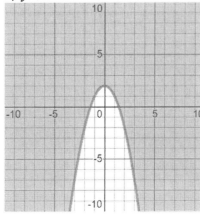

6) $y \leq x^2 - 2x - 3$

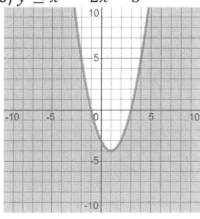

SIFT Test Review

The Selection Instrument for Flight Training (known as SIFT) is used to select applicants for Army Aviation career. The SIFT test is a multiple-aptitude battery that measures developed abilities and helps predict future academic and occupational success in the military. You can only take the SIFT test twice in your lifetime. The SIFT was introduced to replace the older AFAST test.

The SIFT is a multiple-aptitude battery that measures developed abilities and helps predict future academic and occupational success in the military. The SIFT is a multiple-choice test which consists of 7 subtests.

- Section 1 – Simple Drawings (SD)
- Section 2 – Hidden Figures (HF)
- Section 3 – Army Aviation Information Test (AAIT)
- Section 4 – Spatial Apperception Test (SAT)
- Section 5 – Reading Comprehension Test (RCT)
- Section 6 – Math Skills Test (MST)
- Section 7 – Mechanical Comprehension Test (MCT)

The first 5 sections of the SIFT (SD, HF, AAIT, SAT, RCT) have a fixed number of questions which must be answered within a fixed period of time. The last 2 SIFT sections (MST and MCT) are computer adaptive. It means that if the correct answer is chosen, the next question will be harder. If the answer given is incorrect, the next question will be easier. This also means that once an answer is selected on the CAT it cannot be changed.

There are about 30 to 40 Math questions on SIFT and you cannot use a calculator, but a few formulas are provided for some questions.

In this book, there are two complete SIFT Math practice tests. Take these tests to see what score you'll be able to receive on a real SIFT Math test.

Good luck!

Time to Test

Time to refine your quantitative reasoning skill with a practice test

Take a SIFT Math test to simulate the test day experience. After you've finished, score your test using the answer keys.

Before You Start

- You'll need a pencil, a calculator and a timer to take the test.
- For each question, there are five possible answers. Choose which one is best.
- It's okay to guess. There is no penalty for wrong answers.
- Use the answer sheet provided to record your answers.
- You have 40 minutes to complete each test. Answer as many questions as possible.
- After you've finished the test, review the answer key to see where you went wrong.

Good Luck!

SIFT Math Practice Test 1

2020 - 2021

Total number of questions: 40

Total time: 40 Minutes

On a real SIFT test, Math Formulas are provided for some questions.

You may NOT use a calculator on this practice test.

SIFT Math Practice Tests Answer Sheet

Remove (or photocopy) this answer sheet and use it to complete the practice tests.

SIFT Math Practice Test Answer Sheet

SIFT Math Practice Test			
1	A B C D E	16	A B C D E	31	A B C D E
2	A B C D E	17	A B C D E	32	A B C D E
3	A B C D E	18	A B C D E	33	A B C D E
4	A B C D E	19	A B C D E	34	A B C D E
5	A B C D E	20	A B C D E	35	A B C D E
6	A B C D E	21	A B C D E	36	A B C D E
7	A B C D E	22	A B C D E	37	A B C D E
8	A B C D E	23	A B C D E	38	A B C D E
9	A B C D E	24	A B C D E	39	A B C D E
10	A B C D E	25	A B C D E	40	A B C D E
11	A B C D E	26	A B C D E
12	A B C D E	27	A B C D E
13	A B C D E	28	A B C D E
14	A B C D E	29	A B C D E
15	A B C D E	30	A B C D E

SIFT Mathematics Formula Sheet

Area of a:

Parallelogram $$A = bh$$

Trapezoid $$A = \frac{1}{2}h(b_1 + b_2)$$

Surface Area and Volume of a:

Rectangular/Right Prism $$SA = ph + 2B \qquad V = Bh$$

Cylinder $$SA = 2\pi rh + 2\pi r^2 \qquad V = \pi r^2 h$$

Pyramid $$SA = \frac{1}{2}ps + B \qquad V = \frac{1}{3}Bh$$

Cone $$SA = \pi r + \pi r^2 \qquad V = \frac{1}{3}\pi r^2 h$$

Sphere $$SA = 4\pi r^2 \qquad V = \frac{4}{3}\pi r^3$$

(p = perimeter of base B; $\pi = 3.14$)

Algebra

Slope of a line $$m = \frac{y_2 - y_1}{x_2 - x_1}$$

Slope-intercept form of the equation of a line $$y = mx + b$$

Point-slope form of the Equation of a line $$y - y_1 = m(x - x_1)$$

Standard form of a Quadratic equation $$y = ax^2 + bx + c$$

Quadratic formula $$x = \frac{-b \pm \sqrt{b^2 - 4ac}}{2a}$$

Pythagorean theorem $$a^2 + b^2 = c^2$$

Simple interest $$I = prt$$
(I = interest, p = principal, r = rate, t = time)

1) In five successive hours, a car traveled $40\ km, 45\ km, 50\ km, 35\ km$ and $55\ km$. In the next five hours, it traveled with an average speed of $65\ km\ per\ hour$. Find the total distance the car traveled in 10 hours.
   A. $425\ km$
   B. $450\ km$
   C. $550\ km$
   D. $600\ km$
   E. $1,000\ km$

2) How long does a 420-miles trip take moving at 65 miles per hour (mph)?
   A. $4\ hours$
   B. $6\ hours\ and\ 24\ minutes$
   C. $8\ hours\ and\ 24\ minutes$
   D. $8\ hours\ and\ 30\ minutes$
   E. $10\ hours\ and\ 30\ minutes$

3) Right triangle ABC has two legs of lengths $5\ cm$ (AB) and $12\ cm$ (AC). What is the length of the third side (BC)?
   A. $4\ cm$
   B. $6\ cm$
   C. $8\ cm$
   D. $13\ cm$
   E. $20\ cm$

4) The ratio of boys to girls in a school is $2:3$. If there are 500 students in a school, how many boys are in the school?
   A. 540
   B. 360
   C. 300
   D. 280
   E. 200

5) $(7x + 2y)(5x + 2y) = ?$
   A. $2x^2 + 14xy + 2y^2$
   B. $2x^2 + 4xy + 2y^2$
   C. $7x^2 + 14xy + y^2$
   D. $10x^2 + 14xy + 4y$
   E. $35x^2 + 24xy + 4y^2$

6) Which of the following expressions is equivalent to $5x(4+2y)$?
   A. $x + 10xy$
   B. $5x + 5xy$
   C. $20xy + 2xy$
   D. $20x + 5xy$
   E. $20x + 10xy$

7) If $y = 5ab + 3b^3$, what is y when $a = 2$ and $b = 3$?
   A. 24
   B. 31
   C. 36
   D. 51
   E. 111

8) 15 is What percent of 20?
   A. 20%
   B. 25%
   C. 75%
   D. 150%
   E. 300%

9) The perimeter of the trapezoid below is 64. What is its area?
   A. $252\ cm^2$
   B. $234\ cm^2$
   C. $216\ cm^2$
   D. $154\ cm^2$
   E. $260\ cm^2$

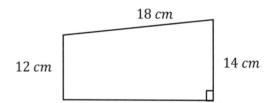

10) Two third of 15 is equal to $\frac{2}{5}$ of what number?
    A. 12
    B. 20
    C. 25
    D. 60
    E. 90

11) The marked price of a computer is D dollar. Its price decreased by 25% in January and later increased by 10% in February. What is the final price of the computer in D dollar?
    A. $0.80\ D$
    B. $0.82\ D$
    C. $0.90\ D$
    D. $1.20\ D$
    E. $1.40\ D$

12) The radius of a cylinder is 8 inches and its height is 14 inches. What is the surface area of the cylinder?
   A. $64 \pi \, in^2$
   B. $128 \pi \, in^2$
   C. $192 \pi in^2$
   D. $256 \pi \, in^2$
   E. $352 \pi in^2$

13) The average of 13, 15, 20 and x is 20. What is the value of x?
   A. 9
   B. 15
   C. 18
   D. 20
   E. 32

14) The price of a sofa is decreased by 25% to $450. What was its original price?
   A. $480
   B. $520
   C. $560
   D. $600
   E. $800

15) The area of a circle is 49π. What is the circumference of the circle?
   A. 7π
   B. 14π
   C. 32π
   D. 64π
   E. 124π

16) A $50 shirt now selling for $28 is discounted by what percent?
   A. 20%
   B. 44%
   C. 54%
   D. 60%
   E. 80%

17) In 1999, the average worker's income increased $2,000 per year starting from $26,000 annual salary. Which equation represents income greater than average? (I = income, x = number of years after 1999)
   A. $I > 2000 \, x + 26000$
   B. $I > -2000 \, x + 26000$
   C. $I < -2000 \, x + 26000$
   D. $I < 2000 \, x - 26000$
   E. $I < 24,000 \, x + 26000$

18) A boat sails 60 miles south and then 80 miles east. How far is the boat from its start point?
   A. 45 miles
   B. 50 miles
   C. 60 miles
   D. 70 miles
   E. 100 miles

19) Sophia purchased a sofa for $530.40. The sofa is regularly priced at $631. What was the percent discount Sophia received on the sofa?
   A. 12%
   B. 16%
   C. 20%
   D. 25%
   E. 40%

20) The score of Emma was half as that of Ava and the score of Mia was twice that of Ava. If the score of Mia was 40, what is the score of Emma?
   A. 10
   B. 15
   C. 20
   D. 30
   E. 40

21) A bag contains 18 balls: two green, five black, eight blue, a brown, a red and one white. If 17 balls are removed from the bag at random, what is the probability that a brown ball has been removed?
   A. $\frac{1}{9}$
   B. $\frac{1}{6}$
   C. $\frac{16}{11}$
   D. $\frac{17}{18}$
   E. $\frac{1}{2}$

22) The average of five consecutive numbers is 36. What is the smallest number?
   A. 38
   B. 36
   C. 34
   D. 12
   E. 8

23) The price of a car was $28,000 in 2012. In 2013, the price of that car was $18,200. What was the rate of depreciation of the price of car per year?
   A. 20%
   B. 30%
   C. 35%
   D. 40%
   E. 50%

24) The width of a box is one third of its length. The height of the box is one third of its width. If the length of the box is 36 cm, what is the volume of the box?
   A. $81\ cm^3$
   B. $162\ cm^3$
   C. $243\ cm^3$
   D. $1,728\ cm^3$
   E. $1,880\ cm^3$

25) A tree 32 feet tall casts a shadow 12 feet long. Jack is 6 feet tall. How long is Jack's shadow?
   A. $2.25\ feet$
   B. $4\ feet$
   C. $4.25\ feet$
   D. $8\ feet$
   E. $12\ feet$

26) When a number is subtracted from 28 and the difference is divided by that number, the result is 3. What is the value of the number?
   A. 2
   B. 4
   C. 7
   D. 12
   E. 24

27) An angle is equal to one ninth of its supplement. What is the measure of that angle?
   A. 9
   B. 18
   C. 25
   D. 60
   E. 90

28) John traveled 150 km in 6 hours and Alice traveled 140 km in 4 hours. What is the ratio of the average speed of John to average speed of Alice?
   A. $3 : 2$
   B. $2 : 3$
   C. $5 : 7$
   D. $5 : 6$
   E. $11 : 16$

29) If 45% of a class are girls, and 25% of girls play tennis, what percent of the class play tennis?
    A. 11%
    B. 15%
    C. 20%
    D. 40%
    E. 80%

30) How many tiles of 8 cm^2 is needed to cover a floor of dimension 7 cm by 24 cm?
    A. 6
    B. 12
    C. 18
    D. 21
    E. 36

31) A rope weighs 600 grams per meter of length. What is the weight in kilograms of 14.2 meters of this rope? (1 $kilograms = 1,000\ grams$)
    A. 0.0852
    B. 0.852
    C. 8.52
    D. 8,520
    E. 85,200

32) A chemical solution contains 6% alcohol. If there is 24 ml of alcohol, what is the volume of the solution?
    A. 240 ml
    B. 400 ml
    C. 600 ml
    D. 1,200 ml
    E. 2,400 ml

33) The average weight of 18 girls in a class is 56 kg and the average weight of 32 boys in the same class is 62 kg. What is the average weight of all the 50 students in that class?
    A. 50
    B. 59.84
    C. 61.68
    D. 61.90
    E. 62.20

34) The price of a laptop is decreased by 20% to $360. What is its original price?
    A. $320
    B. $380
    C. $400
    D. $450
    E. $500

35) A bank is offering 4.5% simple interest on a savings account. If you deposit $9,000, how much interest will you earn in five years?
   A. $360
   B. $720
   C. $2,025
   D. $3,600
   E. $4,800

36) Multiply and write the product in scientific notation:
$$(2.9 \times 10^6) \times (2.6 \times 10^{-5})$$
   A. 754×100
   B. 75.4×10^6
   C. 75.4×10^{-5}
   D. 7.54×10^{11}
   E. 7.54×10

37) If the height of a right pyramid is $14\ cm$ and its base is a square with side $6\ cm$. What is its volume?
   A. $432\ cm^3$
   B. $3088\ cm^3$
   C. $236\ cm^3$
   D. $172\ cm^3$
   E. $168\ cm^3$

38) 5 less than twice a positive integer is 73. What is the integer?
   A. 39
   B. 41
   C. 42
   D. 44
   E. 50

39) A shirt costing $300 is discounted 15%. After a month, the shirt is discounted another 15%. Which of the following expressions can be used to find the selling price of the shirt?
   A. $(300)(0.70)$
   B. $(300) - 300(0.30)$
   C. $(300)(0.15) - (300)(0.15)$
   D. $(300)(0.85)(0.85)$
   E. $(300)(0.85)(0.85) - (300)(0.15)$

40) The sum of six different negative integers is −80. If the smallest of these integers is −16, what is the largest possible value of one of the other five integers?

　　A. −14
　　B. −11
　　C. −10
　　D. −1
　　E. 1

End of SIFT Math Practice Test

SIFT Math
Practice Test 2

2020 - 2021

Total number of questions: 40

Total time: 40 Minutes

On a real SIFT test, Math Formulas are provided for some questions.

You may NOT use a calculator on this practice test.

SIFT Math Practice Tests Answer Sheet

Remove (or photocopy) this answer sheet and use it to complete the practice tests.

SIFT Math Practice Test Answer Sheet

#		#		#	
1	Ⓐ Ⓑ Ⓒ Ⓓ Ⓔ	16	Ⓐ Ⓑ Ⓒ Ⓓ Ⓔ	31	Ⓐ Ⓑ Ⓒ Ⓓ Ⓔ
2	Ⓐ Ⓑ Ⓒ Ⓓ Ⓔ	17	Ⓐ Ⓑ Ⓒ Ⓓ Ⓔ	32	Ⓐ Ⓑ Ⓒ Ⓓ Ⓔ
3	Ⓐ Ⓑ Ⓒ Ⓓ Ⓔ	18	Ⓐ Ⓑ Ⓒ Ⓓ Ⓔ	33	Ⓐ Ⓑ Ⓒ Ⓓ Ⓔ
4	Ⓐ Ⓑ Ⓒ Ⓓ Ⓔ	19	Ⓐ Ⓑ Ⓒ Ⓓ Ⓔ	34	Ⓐ Ⓑ Ⓒ Ⓓ Ⓔ
5	Ⓐ Ⓑ Ⓒ Ⓓ Ⓔ	20	Ⓐ Ⓑ Ⓒ Ⓓ Ⓔ	35	Ⓐ Ⓑ Ⓒ Ⓓ Ⓔ
6	Ⓐ Ⓑ Ⓒ Ⓓ Ⓔ	21	Ⓐ Ⓑ Ⓒ Ⓓ Ⓔ	36	Ⓐ Ⓑ Ⓒ Ⓓ Ⓔ
7	Ⓐ Ⓑ Ⓒ Ⓓ Ⓔ	22	Ⓐ Ⓑ Ⓒ Ⓓ Ⓔ	37	Ⓐ Ⓑ Ⓒ Ⓓ Ⓔ
8	Ⓐ Ⓑ Ⓒ Ⓓ Ⓔ	23	Ⓐ Ⓑ Ⓒ Ⓓ Ⓔ	38	Ⓐ Ⓑ Ⓒ Ⓓ Ⓔ
9	Ⓐ Ⓑ Ⓒ Ⓓ Ⓔ	24	Ⓐ Ⓑ Ⓒ Ⓓ Ⓔ	39	Ⓐ Ⓑ Ⓒ Ⓓ Ⓔ
10	Ⓐ Ⓑ Ⓒ Ⓓ Ⓔ	25	Ⓐ Ⓑ Ⓒ Ⓓ Ⓔ	40	Ⓐ Ⓑ Ⓒ Ⓓ Ⓔ
11	Ⓐ Ⓑ Ⓒ Ⓓ Ⓔ	26	Ⓐ Ⓑ Ⓒ Ⓓ Ⓔ		
12	Ⓐ Ⓑ Ⓒ Ⓓ Ⓔ	27	Ⓐ Ⓑ Ⓒ Ⓓ Ⓔ		
13	Ⓐ Ⓑ Ⓒ Ⓓ Ⓔ	28	Ⓐ Ⓑ Ⓒ Ⓓ Ⓔ		
14	Ⓐ Ⓑ Ⓒ Ⓓ Ⓔ	29	Ⓐ Ⓑ Ⓒ Ⓓ Ⓔ		
15	Ⓐ Ⓑ Ⓒ Ⓓ Ⓔ	30	Ⓐ Ⓑ Ⓒ Ⓓ Ⓔ		

SIFT Math Practice Test

SIFT Mathematics Formula Sheet

Area of a:

Parallelogram $$A = bh$$

Trapezoid $$A = \frac{1}{2}h(b_1 + b_2)$$

Surface Area and Volume of a:

Rectangular/Right Prism $\qquad SA = ph + 2B \qquad V = Bh$

Cylinder $\qquad SA = 2\pi rh + 2\pi r^2 \qquad V = \pi r^2 h$

Pyramid $\qquad SA = \frac{1}{2}ps + B \qquad V = \frac{1}{3}Bh$

Cone $\qquad SA = \pi r + \pi r^2 \qquad V = \frac{1}{3}\pi r^2 h$

Sphere $\qquad SA = 4\pi r^2 \qquad V = \frac{4}{3}\pi r^3$

(p = perimeter of base B; $\pi = 3.14$)

Algebra

Slope of a line $$m = \frac{y_2 - y_1}{x_2 - x_1}$$

Slope-intercept form of the equation of a line $$y = mx + b$$

Point-slope form of the Equation of a line $$y - y_1 = m(x - x_1)$$

Standard form of a Quadratic equation $$y = ax^2 + bx + c$$

Quadratic formula $$x = \frac{-b \pm \sqrt{b^2 - 4ac}}{2a}$$

Pythagorean theorem $$a^2 + b^2 = c^2$$

Simple interest $$I = prt$$
(I = interest, p = principal, r = rate, t = time)

1) When a number is subtracted from 24 and the difference is divided by that number, the result is 3. What is the value of the number?
   A. 2
   B. 4
   C. 6
   D. 12
   E. 24

2) An angle is equal to one fifth of its supplement. What is the measure of that angle?
   A. 20
   B. 30
   C. 45
   D. 60
   E. 90

3) John traveled 150 km in 6 hours and Alice traveled 180 km in 4 hours. What is the ratio of the average speed of John to average speed of Alice?
   A. 3 : 2
   B. 2 : 3
   C. 5 : 9
   D. 5 : 6
   E. 11 : 16

4) If 40% of a class are girls, and 35% of girls play tennis, what percent of the class play tennis?
   A. 10%
   B. 14%
   C. 20%
   D. 40%
   E. 80%

5) In five successive hours, a car traveled 40 km, 45 km, 50 km, 35 km and 55 km. In the next five hours, it traveled with an average speed of 50 $km\ per\ hour$. Find the total distance the car traveled in 10 hours.
   A. 425 km
   B. 450 km
   C. 475 km
   D. 500 km
   E. 1,000 km

6) How long does a 420-miles trip take moving at 50 miles per hour (mph)?
   A. 4 hours
   B. 6 hours and 24 minutes
   C. 8 hours and 24 minutes
   D. 8 hours and 30 minutes
   E. 10 hours and 30 minutes

7) Right triangle ABC has two legs of lengths 6 cm (AB) and 8 cm (AC). What is the length of the third side (BC)?
   A. 4 cm
   B. 6 cm
   C. 8 cm
   D. 10 cm
   E. 20 cm

8) The ratio of boys to girls in a school is 2: 3. If there are 600 students in a school, how many boys are in the school.
   A. 540
   B. 360
   C. 300
   D. 280
   E. 240

9) 25 is What percent of 20?
   A. 20%
   B. 25%
   C. 125%
   D. 150%
   E. 300%

10) The perimeter of the trapezoid below is 54. What is its area?
    A. 252 cm^2
    B. 234 cm^2
    C. 216 cm^2
    D. 154 cm^2
    E. 130 cm^2

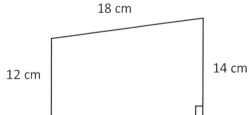

11) Two third of 18 is equal to $\frac{2}{5}$ of what number?
    A. 12
    B. 20
    C. 30
    D. 60
    E. 90

12) The marked price of a computer is D dollar. Its price decreased by 20% in January and later increased by 10% in February. What is the final price of the computer in D dollar?
    A. $0.80\ D$
    B. $0.88\ D$
    C. $0.90\ D$
    D. $1.20\ D$
    E. $1.40\ D$

13) The area of a circle is 25π. What is the circumference of the circle?
    A. 5π
    B. 10π
    C. 32π
    D. 64π
    E. 124π

14) In 1999, the average worker's income increased $3,000 per year starting from $24,000 annual salary. Which equation represents income greater than average? (I = income, x = number of years after 1999)
    A. $I > 3000\ x + 24000$
    B. $I > -3000\ x + 24000$
    C. $I < -3000\ x + 24000$
    D. $I < 3000\ x - 24000$
    E. $I < 24,000\ x + 24000$

15) From last year, the price of gasoline has increased from $1.25 per gallon to $1.75 per gallon. The new price is what percent of the original price?
    A. 72%
    B. 120%
    C. 140%
    D. 160%
    E. 180%

16) A boat sails 40 miles south and then 30 miles east. How far is the boat from its start point?
   A. 45 miles
   B. 50 miles
   C. 60 miles
   D. 70 miles
   E. 80 miles

17) Sophia purchased a sofa for $530.40. The sofa is regularly priced at $624. What was the percent discount Sophia received on the sofa?
   A. 12%
   B. 15%
   C. 20%
   D. 25%
   E. 40%

18) The score of Emma was half as that of Ava and the score of Mia was twice that of Ava. If the score of Mia was 60, what is the score of Emma?
   A. 12
   B. 15
   C. 20
   D. 30
   E. 40

19) The average of five consecutive numbers is 38. What is the smallest number?
   A. 38
   B. 36
   C. 34
   D. 12
   E. 8

20) How many tiles of $8\ cm^2$ is needed to cover a floor of dimension $6\ cm$ by $24\ cm$?
   A. 6
   B. 12
   C. 18
   D. 24
   E. 36

21) A rope weighs 600 grams per meter of length. What is the weight in kilograms of 12.2 meters of this rope? ($1\ kilograms = 1000\ grams$)
   A. 0.0732
   B. 0.732
   C. 7.32
   D. 7,320
   E. 73,200

22) A chemical solution contains 4% alcohol. If there is 24 ml of alcohol, what is the volume of the solution?
   A. 240 ml
   B. 480 ml
   C. 600 ml
   D. 1,200 ml
   E. 2,400 ml

23) The average weight of 18 girls in a class is 60 kg and the average weight of 32 boys in the same class is 62 kg. What is the average weight of all the 50 students in that class?
   A. 60
   B. 61.28
   C. 61.68
   D. 61.90
   E. 62.20

24) The price of a laptop is decreased by 10% to $360. What is its original price?
   A. $320
   B. $380
   C. $400
   D. $450
   E. $500

25) The radius of a cylinder is 8 inches and its height is 12 inches. What is the surface area of the cylinder?
   A. $64 \pi \ in^2$
   B. $128 \pi \ in^2$
   C. $192 \pi \ in^2$
   D. $256 \pi \ in^2$
   E. $320 \pi \ in^2$

26) The average of 13, 15, 20 and x is 18. What is the value of x?
   A. 9
   B. 15
   C. 18
   D. 20
   E. 24

27) The price of a sofa is decreased by 25% to $420. What was its original price?
   A. $480
   B. $520
   C. $560
   D. $600
   E. $800

28) A bank is offering 4.5% simple interest on a savings account. If you deposit $8,000, how much interest will you earn in five years?
   A. $360
   B. $720
   C. $1,800
   D. $3,600
   E. $4,800

29) Multiply and write the product in scientific notation:

$$(4.2 \times 10^6) \times (2.6 \times 10^{-5})$$

   A. 1092×10
   B. 10.92×10^6
   C. 109.2×10^{-5}
   D. 10.92×10^{11}
   E. 1.092×10^2

30) If the height of a right pyramid is $12\ cm$ and its base is a square with side $6\ cm$. What is its volume?
   A. $32\ cm^3$
   B. $36\ cm^3$
   C. $48\ cm^3$
   D. $72\ cm^3$
   E. $144\ cm^3$

31) Solve for x: $4(x + 1) = 6(x - 4) + 20$
   A. 12
   B. 8
   C. 6.2
   D. 5.5
   E. 4

32) Which of the following expressions is equivalent to

$$2x(4 + 2y)?$$

   A. $2xy + 8x$
   B. $8xy + 8x$
   C. $xy + 8$
   D. $2xy + 8x$
   E. $4xy + 8x$

33) If $y = 4ab + 3b^3$, what is y when $a = 2$ and $b = 3$?
   A. 24
   B. 31
   C. 36
   D. 51
   E. 105

34) 11 yards 6 feet and 4 inches equals to how many inches?
   A. 388
   B. 468
   C. 472
   D. 476
   E. 486

35) 5 less than twice a positive integer is 83. What is the integer?
   A. 39
   B. 41
   C. 42
   D. 44
   E. 50

36) A shirt costing $200 is discounted 15%. After a month, the shirt is discounted another 15%. Which of the following expressions can be used to find the selling price of the shirt?
   A. $(200)(0.70)$
   B. $(200) - 200(0.30)$
   C. $(200)(0.15) - (200)(0.15)$
   D. $(200)(0.85)(0.85)$
   E. $(200)(0.85)(0.85) - (200)(0.15)$

37) Which of the following points lies on the line $2x + 4y = 10$
   A. $(2, 1)$
   B. $(-1, 3)$
   C. $(-2, 2)$
   D. $(2, 2)$
   E. $(2, 8)$

38) The price of a car was $20,000 in 2014, $16,000 in 2015 and $12,800 in 2016. What is the rate of depreciation of the price of car per year?
   A. 15%
   B. 20%
   C. 25%
   D. 30%
   E. 50%

39) A ladder leans against a wall forming a 60° angle between the ground and the ladder. If the bottom of the ladder is 30 feet away from the wall, how long is the ladder?
   A. 30 feet
   B. 40 feet
   C. 50 feet
   D. 60 feet
   E. 120 feet

40) Right triangle ABC is shown below. Which of the following is true for all possible values of angle A and B?

   A. $A° + C° = B°$
   B. $A° + B° = 90°$
   C. $A° + C° + B° = 360$
   D. $B° + A° > 90$
   E. $A° + B° < 90°$

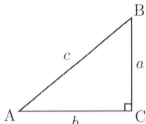

End of SIFT Math Practice Test

SIFT Math Practice Tests
Answer Keys

Now, it's time to review your results to see where you went wrong and what areas you need to improve.

SIFT Math Practice Test 1				SIFT Math Practice Test 2			
1	C	21	D	1	C	21	C
2	B	22	C	2	B	22	C
3	D	23	C	3	C	23	B
4	E	24	D	4	B	24	C
5	E	25	A	5	C	25	E
6	E	26	C	6	C	26	E
7	E	27	B	7	D	27	C
8	C	28	C	8	E	28	C
9	E	29	A	9	C	29	E
10	C	30	D	10	E	30	E
11	B	31	C	11	C	31	E
12	E	32	B	12	B	32	E
13	E	33	B	13	B	33	E
14	D	34	D	14	A	34	C
15	B	35	C	15	C	35	D
16	B	36	E	16	B	36	D
17	A	37	E	17	B	37	B
18	E	38	A	18	B	38	B
19	B	39	D	19	B	39	D
20	A	40	C	20	C	40	B

SIFT Math Practice Tests

Answers and Explanations

SIFT Math Practice Test 1

1) Choice C is correct

Add the first 5 numbers. $40 + 45 + 50 + 35 + 55 = 225$, To find the distance traveled in the next 5 hours, multiply the average by number of hours. $Distance = Average \times Rate = 65 \times 5 = 325$. Add both numbers. $325 + 225 = 550$

2) Choice B is correct

Use distance formula: $Distance = Rate \times time \Rightarrow 420 = 65 \times T$, divide both sides by 65. $420 \div 65 = T \Rightarrow T = 6.4\ hours$. Change hours to minutes for the decimal part. $0.4\ hours = 0.4 \times 60 = 24\ minutes$.

3) Choice D is correct

Use Pythagorean Theorem: $a^2 + b^2 = c^2 \Rightarrow 5^2 + 12^2 = c^2 \Rightarrow 169 = c^2 \Rightarrow c = 13$

4) Choice E is correct

Th ratio of boy to girls is $2:3$. Therefore, there are 2 boys out of 5 students. To find the answer, first divide the total number of students by 5, then multiply the result by 2.

$500 \div 5 = 100 \Rightarrow 100 \times 2 = 200$

5) Choice E is correct

Use FOIL (First, Out, In, Last). $(7x + 2y)(5x + 2y) = 35x^2 + 14xy + 10xy + 4y^2 =$

$$35x^2 + 24xy + 4y^2$$

6) Choice E is correct

Use distributive property: $5x(4 + 2y) = 20x + 10xy$

7) Choice E is correct

$y = 5ab + 3b^3$. Plug in the values of a and b in the equation: $a = 2$ and $b = 3$.

$y = 5\ (2)(3) + 3\ (3)^3 = 30 + 3(27) = 30 + 81 = 111$

8) Choice C is correct

$x = \frac{15}{20} = 0.75 = 75\%$.

9) Choice E is correct

The perimeter of the trapezoid is 64. Therefore, the missing side (height) is

www.EffortlessMath.com

$= 64 - 18 - 12 - 14 = 20$. Area of the trapezoid: $A = \frac{1}{2} h (b_1 + b_2) =$

$$\frac{1}{2} (20)(12 + 14) = 260$$

10) Choice C is correct

Let x be the number. Write the equation and solve for x. $\frac{2}{3} \times 15 = \frac{2}{5} \cdot x \Rightarrow \frac{2 \times 15}{3} = \frac{2x}{5}$, use cross multiplication to solve for x. $5 \times 30 = 2x \times 3 \Rightarrow 150 = 6x \Rightarrow x = 25$

11) Choice B is correct

To find the discount, multiply the number by $(100\% - rate\ of\ discount)$.

Therefore, for the first discount we get: $(D)(100\% - 25\%) = (D)(0.75) = 0.75\ D$

For increase of 10%: $(0.75\ D)(100\% + 10\%) = (0.75\ D)(1.10) = 0.82\ D = 82\%\ of\ D$

12) Choice E is correct

Surface Area of a cylinder $= 2\pi r (r + h)$, The radius of the cylinder is 8 inches and its height is 14 inches. Surface Area of a cylinder $= 2 (\pi)(8)(8 + 14) = 352\ \pi$

13) Choice E is correct

$$\text{average} = \frac{\text{sum of terms}}{\text{number of terms}} \Rightarrow 20 = \frac{13 + 15 + 20 + x}{4} \Rightarrow 80 = 48 + x \Rightarrow x = 32$$

14) Choice D is correct

Let x be the original price. If the price of the sofa is decreased by 25% to $450, then: $75\%\ of\ x = 450 \Rightarrow 0.75x = 450 \Rightarrow x = 450 \div 0.75 = 600$

15) Choice B is correct

Use the formula of areas of circles. $Area = \pi r^2 \Rightarrow 49\pi = \pi r^2 \Rightarrow 49 = r^2 \Rightarrow r = 7$

Radius of the circle is 7. Now, use the circumference formula: Circumference $= 2\pi r = 2\pi (7) = 14\ \pi$.

16) Choice B is correct

Use the formula for Percent of Change. $\frac{\text{New Value} - \text{Old Value}}{\text{Old Value}} \times 100\%$.

$\frac{28 - 50}{50} \times 100\% = -44\%$ (negative sign here means that the new price is less than old price).

17) Choice A is correct

Let x be the number of years. Therefore, $2,000 per year equals $2000x$. starting from $26,000 annual salary means you should add that amount to $2000x$. Income more than that is:

$I > 2000x + 26000$

18) Choice E is correct

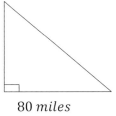

Use the information provided in the question to draw the shape.

Use Pythagorean Theorem: $a^2 + b^2 = c^2$

$60^2 + 80^2 = c^2 \Rightarrow 3600 + 6400 = c^2 \Rightarrow 10000 = c^2 \Rightarrow c = 100$

19) Choice B is correct

The question is this: 530.40 is what percent of 631?

$percent = \frac{530.40}{631} = 84.05 \cong 84$. 530.40 is 84% of 631. Therefore, the discount is:

$100\% - 84\% = 16\%$

20) Choice A is correct

If the score of Mia was 40, therefore the score of Ava is 20. Since, the score of Emma was half as that of Ava, therefore, the score of Emma is 10.

21) Choice D is correct

If 17 balls are removed from the bag at random, there will be one ball in the bag. The probability of choosing a brown ball is 1 out of 18. Therefore, the probability of not choosing a brown ball is 17 out of 18 and the probability of having not a brown ball after removing 17 balls is the same.

22) Choice C is correct

Let x be the smallest number. Then, these are the numbers: $x, x+1, x+2, x+3, x+4$

average $= \frac{\text{sum of terms}}{\text{number of terms}} \Rightarrow 36 = \frac{x+(x+1)+(x+2)+(x+3)+(x+4)}{5} \Rightarrow 36 = \frac{5x+10}{5} \Rightarrow$

$180 = 5x + 10 \Rightarrow 170 = 5x \Rightarrow x = 34$

23) Choice C is correct

Use this formula: Percent of Change: $\frac{\text{New Value} - \text{Old Value}}{\text{Old Value}} \times 100\%$.

$\frac{18{,}200 - 28{,}000}{28{,}000} \times 100\% = -35\%$. The negative sign means that the price decreased

24) Choice D is correct

If the length of the box is 36, then the width of the box is one third of it, 12, and the height of the box is 4 (one third of the width). The volume of the box is: $V = lwh = (36)(12)(4) = 1{,}728$

25) Choice A is correct

Write a proportion and solve for the missing number. $\frac{32}{12} = \frac{6}{x} \rightarrow 32x = 6 \times 12 = 72$

$$32x = 72 \rightarrow x = \frac{72}{32} = 2.25$$

26) Choice C is correct

Let x be the number. Write the equation and solve for x. $(28 - x) \div x = 3$

Multiply both sides by x. $(28 - x) = 3x$, then add x both sides. $28 = 4x$, now divide both sides by 4. $x = 7$

27) Choice B is correct

The sum of supplement angles is 180. Let x be that angle. Therefore, $x + 9x = 180$

$10x = 180$, divide both sides by 10: $x = 18$

28) Choice C is correct

The average speed of john is: $150 \div 6 = 25$, The average speed of Alice is: $140 \div 4 = 35$

Write the ratio and simplify. $25 : 35 \Rightarrow 5 : 7$

29) Choice A is correct

The percent of girls playing tennis is: $45\% \times 25\% = 0.45 \times 0.25 = 0.11 = 11\%$

30) Choice D is correct

The area of the floor is: $7\ cm \times 24\ cm = 168\ cm^2$, The number is tiles needed =

$168 \div 8 = 21$.

31) Choice C is correct

The weight of 14.2 meters of this rope is: $14.2 \times 600\ g = 8520\ g$

$1\ kg = 1000\ g$, therefore, $8520\ g \div 1,000 = 8.52\ kg$

32) Choice B is correct

6% of the volume of the solution is alcohol. Let x be the volume of the solution.

Then: $6\%\ of\ x = 24\ ml \Rightarrow 0.06\ x = 24 \Rightarrow x = 24 \div 0.06 = 400$

33) Choice B is correct

$average = \frac{sum\ of\ terms}{number\ of\ terms}$. The sum of the weight of all girls is: $18 \times 56 = 1,008\ kg$

The sum of the weight of all boys is: $32 \times 62 = 1,984\ kg$. The sum of the weight of all students is: $1,008 + 1,984 = 2,992\ kg$. $average = \frac{2992}{50} = 59.84$

34) Choice D is correct

Let x be the original price. If the price of a laptop is decreased by 20% to \$360, then: $80\%\ of$

$x = 360 \Rightarrow 0.80x = 360 \Rightarrow x = 360 \div 0.80 = 450$

35) Choice C is correct

Use simple interest formula: $I = prt$. (I = interest, p = principal, r = rate, t = time). $I = (9,000)(0.045)(5) = 2,025$

36) Choice E is correct

$(2.9 \times 10^6) \times (2.6 \times 10^{-5}) = (2.9 \times 2.6) \times (10^6 \times 10^{-5}) = 7.54 \times (10^{6+(-5)})$
$= 7.54 \times 10^1$

37) Choice E is correct

The formula of the volume of pyramid is: $V = \frac{l \times w \times h}{3}$

The length and width of the pyramid is 6 cm and its height is 14 cm. Therefore:

$$V = \frac{6 \times 6 \times 14}{3} = 168 \, cm^3$$

38) Choice A is correct

Let x be the integer. Then: $2x - 5 = 73$, Add 5 both sides: $2x = 78$, Divide both sides by 2:

$$x = 39$$

39) Choice D is correct

To find the discount, multiply the number by $(100\% - ate\ of\ discount)$. Therefore, for the first discount we get: $(300)(100\% - 15\%) = (300)(0.85)$. For the next 15% discount: $(300)(0.85)(0.85)$.

40) Choice C is correct

The smallest number is -15. To find the largest possible value of one of the other five integers, we need to choose the smallest possible integers for four of them. Let x be the largest number. Then: $-80 = (-16) + (-15) + (-14) + (-13) + (-13) + x \rightarrow -80 = -70 + x, \rightarrow$

$$x = -80 + 70 = -10$$

SIFT Math Practice Test 2

1) Choice C is correct

Let x be the number. Write the equation and solve for x. $(24 - x) \div x = 3$. Multiply both sides by x. $(24 - x) = 3x$, then add x both sides. $24 = 4x$, now divide both sides by 4.

$x = 6$

2) Choice B is correct

The sum of supplement angles is 180. Let x be that angle. Therefore, $x + 5x = 180$

$6x = 180$, divide both sides by 6: $x = 30$

3) Choice C is correct

The average speed of john is: $150 \div 6 = 25$, The average speed of Alice is: $180 \div 4 = 45$

Write the ratio and simplify. $25:45 \Rightarrow 5:9$

4) Choice B is correct

The percent of girls playing tennis is: $40\% \times 35\% = 0.40 \times 0.35 = 0.14 = 14\%$

5) Choice C is correct

Add the first 5 numbers. $40 + 45 + 50 + 35 + 55 = 225$

To find the distance traveled in the next 5 hours, multiply the average by number of hours.

$Distance = Average \times Rate = 50 \times 5 = 250$, Add both numbers. $250 + 225 = 475$

6) Choice C is correct

Use distance formula: $Distance = Rate \times time \Rightarrow 420 = 50 \times T$, divide both sides by 50. $420 \div 50 = T \Rightarrow T = 8.4\ hours$. Change hours to minutes for the decimal part. $0.4\ hours = 0.4 \times 60 = 24\ minutes$.

7) Choice D is correct

Use Pythagorean Theorem: $a^2 + b^2 = c^2$, $6^2 + 8^2 = c^2 \Rightarrow 100 = c^2 \Rightarrow c = 10$

8) Choice E is correct

Th ratio of boy to girls is $2:3$. Therefore, there are 2 boys out of 5 students. To find the answer, first divide the total number of students by 5, then multiply the result by 2.

$600 \div 5 = 120 \Rightarrow 120 \times 2 = 240$

9) Choice C is correct

Use percent formula: $part = \frac{percent}{100} \times whole$

$$25 = \frac{percent}{100} \times 20 \Rightarrow 25 = \frac{percent \times 20}{100} \Rightarrow 25 = \frac{percent \times 2}{10}, multiply\ both\ sides\ by\ 10.$$

$250 = percent \times 2$, divide both sides by 2. $125 = percent$

10) Choice E is correct

The perimeter of the trapezoid is 54.

Therefore, the missing side (height) is $= 54 - 18 - 12 - 14 = 10$

Area of the trapezoid: $A = \frac{1}{2} h (b_1 + b_2) = \frac{1}{2} (10) (12 + 14) = 130$

11) Choice C is correct

Let x be the number. Write the equation and solve for x.

$\frac{2}{3} \times 18 = \frac{2}{5} \cdot x \Rightarrow \frac{2 \times 18}{3} = \frac{2x}{5}$, use cross multiplication to solve for x.

$5 \times 36 = 2x \times 3 \Rightarrow 180 = 6x \Rightarrow x = 30$

12) Choice B is correct

To find the discount, multiply the number by $(100\% - rate\ of\ discount)$.

Therefore, for the first discount we get: $(D)(100\% - 20\%) = (D)(0.80) = 0.80\ D$

For increase of 10%: $(0.80\ D)(100\% + 10\%) = (0.80\ D)(1.10) = 0.88\ D = 88\%\ of\ D$

13) Choice B is correct

Use the formula of areas of circles. $Area = \pi r^2 \Rightarrow 25\ \pi = \pi r^2 \Rightarrow 25 = r^2 \Rightarrow r = 5$

Radius of the circle is 5. Now, use the circumference formula: Circumference $= 2\pi r = 2\pi (5) = 10\ \pi$

14) Choice A is correct

Let x be the number of years. Therefore, $3,000 per year equals $2000x$. starting from $24,000 annual salary means you should add that amount to $3000x$. Income more than that is:

$I > 3000\ x + 24000$

15) Choice C is correct

The question is this: 1.75 is what percent of 1.25? Use percent formula:

$part = \frac{percent}{100} \times whole$

$$1.75 = \frac{percent}{100} \times 1.25 \Rightarrow 1.75 = \frac{percent \times 1.25}{100} \Rightarrow 175 = percent \times 1.25$$

$$\Rightarrow percent = \frac{175}{1.25} = 140$$

16) Choice B is correct

Use the information provided in the question to draw the shape.

Use Pythagorean Theorem: $a^2 + b^2 = c^2$

$40^2 + 30^2 = c^2 \Rightarrow 1600 + 900 = c^2 \Rightarrow 2500 = c^2 \Rightarrow c = 50$

17) Choice B is correct

The question is this: 530.40 is what percent of 624?

Use percent formula: $part = \frac{percent}{100} \times whole$

$$530.40 = \frac{percent}{100} \times 624 \Rightarrow 530.40 = \frac{percent \times 624}{100} \Rightarrow 53040 = percent \times 624 \Rightarrow$$

$$percent = \frac{53040}{624} = 85$$

530.40 is 85% of 624. Therefore, the discount is: $100\% - 85\% = 15\%$

18) Choice B is correct

If the score of Mia was 60, therefore the score of Ava is 30. Since, the score of Emma was half as that of Ava, therefore, the score of Emma is 15.

19) Choice B is correct

Let x be the smallest number. Then, these are the numbers: $x, x + 1, x + 2, x + 3, x + 4$

$average = \frac{sum\ of\ terms}{number\ of\ terms} \Rightarrow 38 = \frac{x+(x+1)+(x+2)+(x+3)+(x+4)}{5} \Rightarrow 38 = \frac{5x+10}{5} \Rightarrow 190 = 5x + 10 \Rightarrow 180 = 5x \Rightarrow x = 36$

20) Choice C is correct

The area of the floor is: $6\ cm \times 24\ cm = 144\ cm^2$, The number is tiles needed $= 144 \div 8 = 18$

21) Choice C is correct

The weight of 12.2 meters of this rope is: $12.2 \times 600\ g = 7320\ g$,

$1\ kg = 1000\ g$, therefore, $7320\ g \div 1000 = 7.32\ kg$

22) Choice C is correct

4% of the volume of the solution is alcohol. Let x be the volume of the solution.

Then: $4\% \ of \ x = 24 \ ml \Rightarrow 0.04 \ x = 24 \Rightarrow x = 24 \div 0.04 = 600$

23) Choice B is correct

$average = \frac{sum \ of \ terms}{number \ of \ terms}$, The sum of the weight of all girls is: $18 \times 60 = 1080 \ kg$, The sum of the weight of all boys is: $32 \times 62 = 1984 \ kg$, The sum of the weight of all students is: $1080 + 1984 = 3064 \ kg$, $average = \frac{3064}{50} = 61.28$

24) Choice C is correct

Let x be the original price. If the price of a laptop is decreased by 10% to $360, then: $90\% \ of \ x = 360 \Rightarrow 0.90x = 360 \Rightarrow x = 360 \div 0.90 = 400$

25) Choice E is correct

Surface Area of a cylinder $= 2\pi r \ (r + h)$, The radius of the cylinder is 8 inches and its height is 12 inches. Surface Area of a cylinder $= 2 \ (\pi) \ (8) \ (8 + 12) = 320 \ \pi$

26) Choice E is correct

$average = \frac{sum \ of \ terms}{number \ of \ terms} \Rightarrow 18 = \frac{13+15+20+x}{4} \Rightarrow 72 = 48 + x \Rightarrow x = 24$

27) Choice C is correct

Let x be the original price. If the price of the sofa is decreased by 25% to $420, then: $75\% \ of \ x = 420 \Rightarrow 0.75x = 420 \Rightarrow x = 420 \div 0.75 = 560$

28) Choice C is correct

Use simple interest formula: $I = prt$, ($I =$ interest, $p =$ principal, $r =$ rate, $t =$ time)

$I = (8,000)(0.045)(5) = 1,800$

29) Choice E is correct

$(4.2 \times 10^6) \times (2.6 \times 10^{-5}) = (4.2 \times 2.6) \times (10^6 \times 10^{-5}) = 10.92 \times (10^{6 + (-5)})$
$= 1.092 \times 10^2$

30) Choice E is correct

The formula of the volume of pyramid is: $V = \frac{l \times w \times h}{3}$. The length and width of the pyramid is 6 cm and its height is 12 cm. Therefore: $V = \frac{6 \times 6 \times 12}{3} = 144 \ cm^3$

31) Choice E is correct

Simplify: $4(x + 1) = 6(x - 4) + 20, 4x + 4 = 6x - 24 + 20, 4x + 4 = 6x - 4$

Subtract $4x$ from both sides: $4 = 2x - 4$, Add 4 to both sides: $8 = 2x, 4 = x$

32) Choice E is correct

Use distributive property: $2x(4 + 2y) = 8x + 4xy = 4xy + 8x$

www.EffortlessMath.com

33) Choice E is correct

$y = 4ab + 3b^3$, plug in the values of a and b in the equation: $a = 2$ and $b = 3$,

$y = 4(2)(3) + 3(3)^3 = 24 + 3(27) = 24 + 81 = 105$

34) Choice C is correct

$11 \times 36 + 6 \times 12 + 4 = 472$

35) Choice D is correct

Let x be the integer. Then: $2x - 5 = 83$, Add 5 both sides: $2x = 88$, Divide both sides by 2: $x = 44$

36) Choice D is correct

To find the discount, multiply the number by $(100\% - rate\ of\ discount)$. Therefore, for the first discount we get: $(200)(100\% - 15\%) = (200)(0.85)$, For the next 15% discount: $(200)(0.85)(0.85)$.

37) Choice B is correct

Plug in each pair of number in the equation:

    A. $(2, 1)$: $2(2) + 4(1) = 8$
    B. $(-1, 3)$: $2(-1) + 4(3) = 10$
    C. $(-2, 2)$: $2(-2) + 4(2) = 4$
    D. $(2, 2)$: $2(2) + 4(2) = 12$

38) Choice B is correct

Use this formula: Percent of Change: $\frac{New\ Value - Old\ Value}{Old\ Value} \times 100\%$

$\frac{16000 - 2000}{20000} \times 100\% = -20\%$ and $\frac{12800 - 16000}{16000} \times 100\% = -20\%$

39) Choice D is correct

The relationship among all sides of special right triangle $30° - 60° - 90°$ is provided in this triangle:

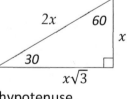

In this triangle, the opposite side of $30°$ angle is half of the hypotenuse.

Draw the shape of this question:

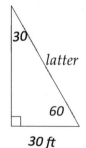

The latter is the hypotenuse. Therefore, the latter is $60\ ft$

40) Choice B is correct.

All angles in a triangle sum up to 180 degrees. From the choices provided, only choice B is correct: $A° + B° = 90°$

www.EffortlessMath.com

... So Much More Online!

✓ FREE Math lessons

✓ More Math learning books!

✓ Mathematics Worksheets

✓ Online Math Tutors

Need a PDF version of this book?

Visit www.EffortlessMath.com

Visit www.EffortlessMath.com
for Online Math Practice

Receive the PDF version of this book or get another FREE book!

Thank you for using our Book!

Do you LOVE this book?

Then, you can get the PDF version of this book or another book absolutely FREE!

Please email us at:

info@EffortlessMath.com

for details.

Made in the USA
Columbia, SC
10 August 2022